T3-BWZ-685

Landscape and People of the Franchthi Region

Excavations at Franchthi Cave, Greece

T. W. Jacobsen, General Editor

FASCICLE 2

Landscape and People of the Franchthi Region

TJEERD H. VAN ANDEL and
SUSAN B. SUTTON

with contributions by
Julie M. Hansen and
Charles J. Vitaliano

INDIANA UNIVERSITY PRESS
Bloomington & Indianapolis

Manufactured in the United States of America

Library of Congress Cataloging-in-Publication Data
Van Andel, Tjeerd H. (Tjeerd Hendrik), 1923-
Landscape and people of the Franchthi region.
(Excavations at Franchthi Cave, Greece ; fasc. 2)
Bibliography: p.
1. Landforms—Greece—Franchthi Cave Site Region.
2. Anthropo-geography—Greece—Franchthi Cave Site Region. 3. Greece—Antiquities. 4. Franchthi Cave Site (Greece) I. Sutton, Susan B. II. Title
III. Series.
GB436.G8V36 1987 938'.8 87-4156
ISBN 0-253-31975-7 (pbk.)
1 2 3 4 5 91 90 89 88 87

CONTENTS

FIGURES

TABLES

PLATES

FOREWORD

This volume is another component of the initial installment of four fascicles in the Franchthi final publication series. Unlike certain other components of the series (e.g., Perlès 1987), it represents something of a compromise in that it offers primary data and levels of interpretation that transcend to a degree our conceptual distinction between ''levels'' of publication in this series (see, e.g., Jacobsen and Farrand 1987). Nevertheless, the nature of its contents is such that its appearance seemed most appropriate at this time and in this form.

The primary objectives of this volume are to provide necessary background for the other studies in the series and to introduce the reader to the study area in which the Franchthi excavations took place. As such, it helps to provide a context, both ancient and modern, for the excavations and resulting publications.

At the same time, this fascicle is valuable as a reminder to the reader of the fundamental relationship between our project and its parent enterprise, the ''Argolid Exploration Project.'' As described more fully elsewhere (Jacobsen, forthcoming), the ''AEP'' was conceived and organized by M. H. Jameson (then of the University of Pennsylvania), later evolved into a collaborative project with Indiana University (with the writer as co-director), and eventually came to be based at Stanford University (hence the occasional references in what follows to the ''Stanford Archaeological Survey''). Both primary authors of this volume are members of that project, and van Andel serves as one of its co-directors. The full publication of the AEP-Stanford Survey is now in its final stages of preparation (Jameson, van Andel, and Runnels, forthcoming; see also van Andel and Runnels 1987), and it will incorporate, complement, and expand upon much of the information provided in this study.

This manuscript was completed by the authors, essentially in its present form, in June, 1986. The general editor would like to express his gratitude once again to Ms. Frances Huber and Ms. Mary Ann Weddle of the Program in Classical Archaeology at Indiana University and to our colleagues at the Indiana University Press, most notably Mr. John Gallman, Ms. Harriet Curry, and Ms. Roberta Diehl, for their considerable assistance and cooperation in the preparation of the manuscript for publication. The photographs in Plates 1-13 were taken by R. Grody, T. W. Jacobsen, S. B. Sutton, and M. and K. Sheehan. As always, we are most grateful to the Indiana University Foundation (E. A. Schrader Endowment for Classical Archaeology) and the National Endowment for the Humanities for their continuing support of our project.

T. W. JACOBSEN

PREFACE

"I have long held the view that our ultimate goal is to determine the inter-relationships between culture and environment."

K. W. Butzer, *Environment and Archaeology* (1964)

Archaeological sites exist within a landscape, a landscape that may have changed substantially over time if it has been occupied long enough or if the inhabitants exerted enough pressure on their surroundings. This, among the many tasks of the archaeological geologist, is the most challenging, the study of a landscape in the context of its occupation and exploitation by human beings. Karl Butzer (1982:1-7) has suggested that the principal objective of any paleoenvironmental study ought to be the array of options presented to its exploiters at various times by a changing landscape. To this I would add the natural complement, the identification of changes in a landscape wrought accidentally or deliberately by its human inhabitants.

The principal options offered by any landscape are food, water, and raw materials, and also, though less often considered, shelter, healthfulness or its opposite, ease of travel, and accessibility from land or sea. The options vary with time and so do the selections human beings make among them. While hunter-gatherers seek what nature offers over large territories, agriculturists prefer landscapes that are easily molded to their needs.

It is obvious that this kind of study requires a broad range of environmental disciplines, including palynology, geology, paleoclimatology, and paleoceanography, as well as the information available from the archaeological sites themselves. In the following pages use will be made of many studies stimulated by the Franchthi excavations directed by T. W. Jacobsen, including what has already been put in print by archaeologists, paleozoologists, and paleoethnobotanists. Key elements also are the areally broader studies of the archaeological survey of the southern Argolid, the Argolid Exploration Project (Jameson et al., forthcoming).

Paleoenvironmental reconstructions, even those done with great skill and based on abundant data, suffer from the difficulty with which confidence limits are established. It would be far simpler to extrapolate backward from the present, assuming that the landscape has varied little and only in a minor way. Unfortunately, even leaving human influence aside, this assumption is not warranted for the past 50,000 years which include the large changes in climate, vegetation, and geography that accompanied the transition from a full glacial to the present late interglacial.

The following discussion of the Franchthi landscape emphasizes the period from ca. 100,000 B.P., long before clear evidence exists for the occupation of the cave, until the site record ends at about 5,000 B.P. The reports of the Argolid Exploration Project focus on the subsequent history with its intense human modification of the landscape (Jameson et al., forthcoming).

In the last two decades our understanding of the events and processes of the Ice Age has undergone profound changes. The CLIMAP Project (CLIMAP 1976) especially has greatly altered our perceptions of the chronology and climate of the late Quaternary. Progress continues apace, and even in the few years that have elapsed since the Argolid field work was finished, our views have already required revision. Moreover, because data regarding past climate and vegetation are sparse for our area, informed speculation looms larger in the following pages than we would have preferred, especially in the summary of the landscape evolution that constitutes the last chapter of Part I.

Another equally important perspective is provided by the study of traditional patterns of living in the area and by the events of its latest history. Greece has until very recently been a conservative society, and many still surviving approaches to economic and social issues have much to teach us regarding the possible cultural, social, and economic patterns in the more remote past. Until the 1940s and 1950s, threshing floors and wooden plows, as well as fallowing were common in agriculture, the construction of wooden boats of time-honored designs can still be observed today, and the construction of houses was, in many ways, not dissimilar to the medieval designs or even those of the later Bronze Age. Thus anthropological studies initiated by the Argolid Exploration Project and summarized in Part II of this work offer much food for contemplation to those who must interpret the prehistoric past.

On the other hand, the archaic nature of technology and customs may deceive and be of much more recent vintage than we, brought up in a technological world, are inclined to assume. One should remember that the present population of the Argolid, in the main, traces its roots no farther than the late first millennium A.D., and that the prehistoric settlers at Franchthi were not its ancestors, biologically or culturally. Rather, it is in the parallel solutions chosen at different times for similar problems in similar environmental settings that the anthropological perspective has its principal value for the archaeologist.

Acknowledgements. As noted above, this synthesis draws heavily on the work, advice, and interest of others. Michael H. Jameson introduced both authors to the area and its potential and has consistently encouraged the landscape study. Thomas W. Jacobsen, director of the Franchthi excavations, not only invited this contribution to the excavation reports but has generously furnished much relevant information and participated in some of its component studies. Data and ideas of various kinds were supplied by our many colleagues, foremost among them Susan Duhon, Hamish Forbes, Julie Hansen, Susan Langdon, Nikolaos Lianos, Sebastian Payne, Catherine Perlès, Kevin O. Pope, Curtis N. Runnels, Judith C. Shackleton, Mark C. Sheehan, and Robert Sutton, and especially William R. Farrand who selflessly reviewed in detail and to its great benefit an early version of this manuscript. The Quaternary geological and marine studies were supported by grants from the National Science Foundation and the National Endowment for the Humanities, by the Underwater Archaeology Branch of the Greek Archaeological Service, and by private donations to Stanford University. Susan Sutton's anthropological research was supported by a National Endowment for the Humanities Summer Stipend. The geological survey and part of J. M. Hansen's work were supported by a grant from the National Endowment for the Humanities to Indiana University, and Hansen and M. C. Sheehan were also partly supported by the E. A. Schrader Endowment at Indiana University. Susan Duhon ably assisted in the geological field work and helped with the assembly of the data for the geological map that appears in Fascicle 1 of this series. [The authors had originally intended that the map should appear in this volume, but, because of its size and production in color, it will appear only in Jacobsen and Farrand (1987: Plate 1). — EDITOR.] The map and the accompanying cross sections were prepared by the combined Indiana Geological Survey and Indiana University Department of Geology drafting staff. A grant from the Dean of Research and Graduate Development, Indiana University, helped defray part of the drafting cost. The Greek Institute of Geology and Mineral Exploration, the Archaeological Service of the Ministry of Culture of the Greek Government, and the Greek National Statistical Service generously gave permission and support to carry out the various investigations. The local officials of the various administrative units of the southern Argolid graciously answered questions and made local records available. Sutton's ethnographic research was greatly aided by the openness and hospitality of residents throughout the area. To all we are deeply indebted.

TJEERD H. VAN ANDEL

PART I

The Landscape

Tjeerd H. van Andel

CHAPTER ONE
The Modern Landscape

PHYSIOGRAPHY AND CLIMATE

Franchthi Cave occupies the southwestern tip of a rounded limestone headland north of Kiladha Bay on the southern Argolid peninsula. The headland is part of a series of low mountain ranges that form the transition between the rugged high country of the central Argolid and the low rolling hills of its far southern tip.

The peninsula, scalloped by numerous bays, is barred in the north by a high east-west range of mountains with Megalovouni (1,113 m) in the center as its highest point. The Dhidhima Range in the west consists of limestone with steep but simple slopes, while the Adheres Range in the east is cut into softer sandstones and shales and hence has a more precipitous and highly dissected relief. Embayed in the south slope of these ranges at an elevation of about 200 m are two plateaus. The Dhidhima plateau is semi-enclosed without external drainage and has at least two large sinkholes, whereas the more open Iliokastro plateau at the foot of Megalovouni is drained by several valleys.

Farther south lie other, lower limestone ridges extending from the Franchthi headland to the Iliokastro plateau. They are separated from the border ranges by the Fourni valley and the headwaters of the Ermioni river above the Katafiki gorge (Figure 1). Another range farther south begins with a series of low hills and north-facing scarps south of Kiladha Bay, trending east and rising in height beyond Kranidhi until it ends in the cape upon which the town of Ermioni stands. South of this ridge the peninsula slopes gently seaward, its rolling hills here and there interrupted by dry, steep-sided valleys.

Many streams drain this rugged country, but their valleys are small and have little bottomland, except for those of the Fourni and Ermioni rivers. The streams are generally dry save after heavy rains; only on the Iliokastro plateau do springs provide water to some upper stream courses as late as August and September.

The coasts are generally steep with cliffs ranging in height from a few to a few tens of meters and are fringed with gravel and cobble beaches. Sand beaches are rare. On the west side, deep, sheltered embayments exist at Kiladha and Porto Kheli, each with a small coastal plain covered with alluvium. Along the east shore one finds several open bays with salt marshes, some open, some barred from the sea by sandy beach ridges. Intervening headlands, such as those of Franchthi or Cape Mouzaki, tend to be steep and rugged and drop directly into the sea. Two semi-enclosed lagoons with muddy bottoms and bordering marshes exist at Ververonda on the west shore and at Thermisi in the northeast. Offshore, but connected to the mainland by a shallow shelf, are the large islands of Idhra, Dhokos, and Spetsai.

In spite of its ruggedness, travel across the southern peninsula is not arduous provided one follows the main transverse stream valleys. To go from Franchthi to Dhidhima or Iliokastro by means of the Fourni and upper Ermioni valleys takes but a few hours on foot.

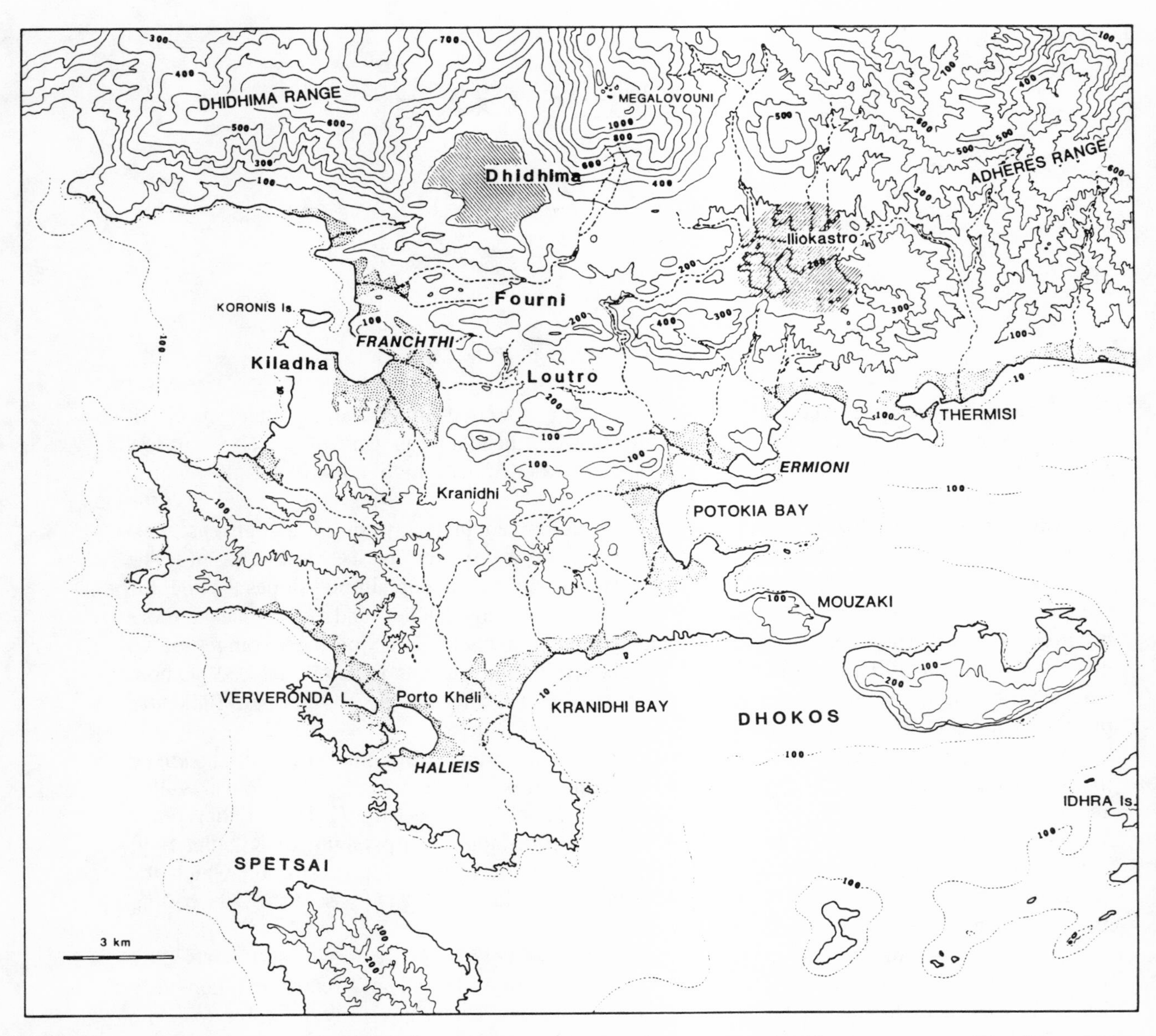

Figure 1. Relief and place names in the southern Argolid. Contour interval 100 m, contours taken from sheets Idhra and Spetsai of the 1:50,000 scale topographic map of Greece issued by the Geographic Division of the Greek Army. Isobaths at 10 m and 100 m are from the same source with modifications after van Andel and Lianos (1983). Principal streams are shown with dashed lines; coastal plains, defined approximately by the 20 m contour, are stippled. Oblique shading indicates the plateaus of Dhidhima and Iliokastro.

One may also pass easily and directly into the Ermioni valley below the Katafiki gorge across the saddle at Loutro, or attain the shore at Ermioni through the valley north of the Kranidhi ridge. Directly towards the south and southeast, travel is less easy because of the many steep stream channels but by no means forbidding.

To go north beyond the southern peninsula, that is, to enter the central and northern Argolid, is another matter. A major range must be crossed to reach the north shore on the Saronic Gulf near well-watered Trizin, for example, and along the western shore the way from Franchthi towards Navplion is barred by steep mountains and precipitous gorges such as the Bedheni valley immediately north of the Dhidhima Range. The best northern route passes Dhidhima, thence climbs across a pass west of Megalovouni into the upland valleys of the Epidhavria. It is not surprising that, until the construction of the motor road not very many years ago, travel in and out of the southern Argolid was almost exclusively by boat.

The climate of the southern Argolid, one of the driest in Greece, is typically Mediterranean with its hot, dry summers and cool, moist winters. This climate is the result of seasonal latitudinal shifts of the zone of global westerlies across central Europe and its interaction with a subtropical high pressure zone lying over northern Africa (Tollner 1976; Perry 1981; Wigley and Farmer 1982). Locally, this regime is altered by orographic factors and by the various basins of the Mediterranean which act as heat reservoirs and sources of moisture.

In winter, the Siberian high-pressure zone reaches southwestward into the Balkans, thus pushing the mid-latitude westerlies southward, only to withdraw in summer when the subtropical high-pressure zone dominates the Mediterranean (Figure 2). The polar jet, close to Greece in the winter, episodically varies its course from zonal to meridional. Zonal flow in the winter brings dry, relatively warm weather, whereas a southerly meridional flow is accompanied by mild, wet storms originating mainly at various points within the Mediterranean itself. A northerly meridional flow brings cooler, moist weather. In the summer, the subtropical high prevents the formation of Mediterranean storms and blocks access for those that originate in the Atlantic. As a result, only occasional summer thunderstorms bring rain during that season.

Because of the large influence of marine and orographic factors, local climatic variations are large. In addition, the climate tends to vary over relatively short periods of time. Large shifts of the northern boundary of the Mediterranean climatic zone have occurred over the past several decades. Between 1940 and 1960, rainfall decreased significantly over the southeastern Peloponnese and increased in central and northern Greece as a result of a southwestward shift of this boundary (Wigley and Farmer 1982). Large annual temperature fluctuations are common.

Climate data for the southern Argolid itself are sparse and must be supplemented with the long record from Navplion at the head of the Gulf of Argos (Philippson 1948; NID 1944). There, the temperature averages 10 °C in January and 27 °C in July (Figure 3), values which H. A. Forbes (1982) regards as reasonable for the southern Argolid also. The year does not really turn hot until late in June, and in September the nights already begin to cool. Frost is uncommon at low elevations even at night (Gavrielides 1976a).

For the period from September, 1971, to September, 1972, judged to be a normal season, Gavrielides (1976a) cites a rainfall of 520 mm in the Fourni valley (Figure 3). The rain fell on 32 separate days. These observations are in good accord with averages at Navplion of 70 mm for January, 5 mm in July, and 495 mm for the entire year.

There are few reports of snow at low elevations in the southern Argolid, although it seems probable that it does occasionally snow above 300 m. On the northern boundary ranges, above 600 m, snow falls several times each winter but does not stay long on the ground.

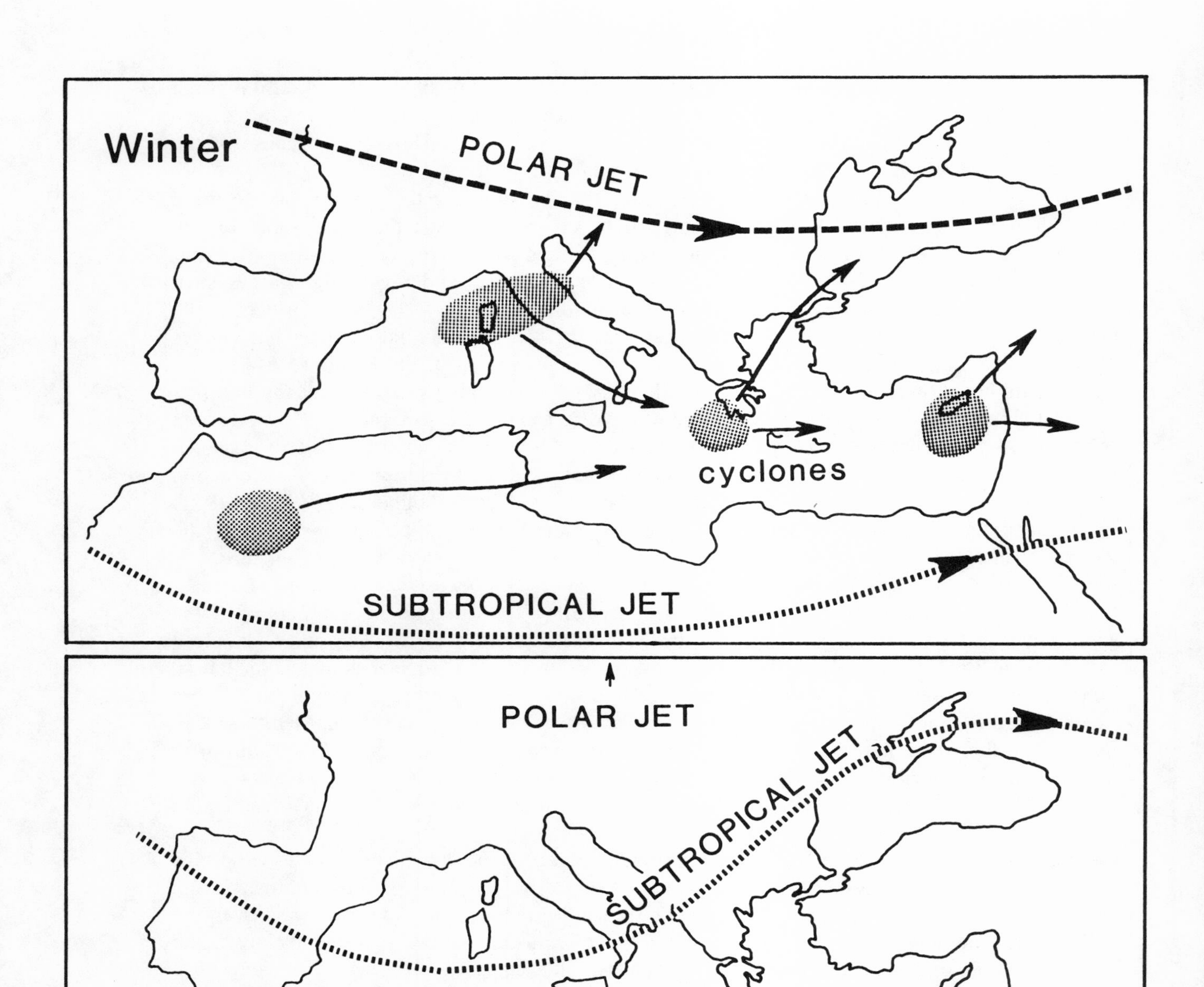

Figure 2. Winter and summer atmospheric circulation and cyclone generation in the Mediterranean. Shaded are areas where rain storms are generated in winter which subsequently travel east and northeast. Modified from Wigley and Farmer (1982:Figs. 1.3 and 1.7).

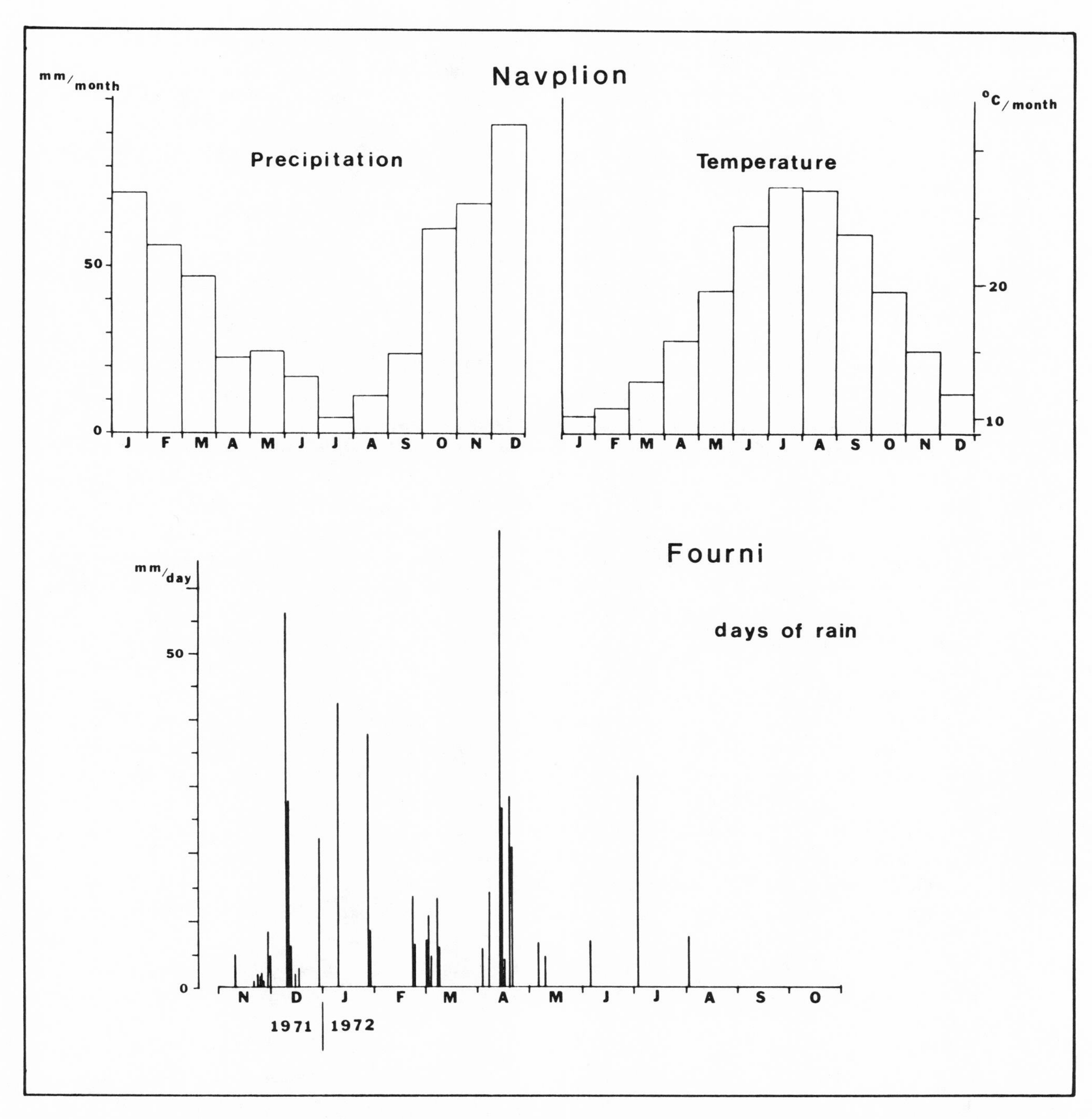

Figure 3. Monthly average temperature and precipitation in the Argolid. *Top:* mean monthly precipitation (*left*) and temperature (*right*) at Navplion on the northern Gulf of Argos (NID 1944:Vol. I). *Bottom:* days and daily amount of rain during the period September, 1971, to September, 1972, at Fourni in the southern Argolid (after Gavrielides 1976a).

In winter, the wind blows mainly from the west or northwest and gales are not uncommon, of which the shores of the Franchthi embayment bear the full brunt. The east side of the peninsula is well protected in that season. The summer brings a daily regime of daytime sea breezes and a nightly landwind, but in late July and August strong "etesian" winds blow from the north or northeast for days on end, often seriously impeding marine traffic.

VEGETATION

The Franchthi region is at present a patchwork of cultivated fields, some temporarily abandoned, interspersed with natural vegetation, and set off by a backdrop of steep mountain slopes covered with maquis or pine woods. The word "natural" is not to be taken literally here; the plant communities of the region are the product of millennia of intense human activity, and the original climax associations of the undisturbed landscape cannot be traced with confidence (Sheehan 1979; Sheehan and Sheehan 1982).

Pollen data (see Chapter Four) indicate that the Middle Holocene plant cover consisted in part of a relatively rich, though probably open deciduous oak wood or parkland. Today, phrygana (garrigue) and maquis are the dominant communities, often regarded as the end products of a progressive degradation of the deciduous oak woods under human influence, mainly by grazing and fuel collecting. They have formed the backdrop of human activity in the area for some 5,000 years or more (Sheehan 1979). Rackham (1983) has argued that, left to themselves, the fallow fields evolve to phrygana and then to maquis, and the maquis to woods dominated by Kermes oak (*Quercus coccifera*) and a variety of other species. Examples of this succession can be seen in the Valley of the Muses in Boeotia and on the island of Samothrace.

Marius (1976; see also Verheye and Lootens-de Muynck 1974; Gavrielides 1976b) has mapped plant communities in the Fourni valley and on the Dhidhima plateau in some detail (Figure 4). Sheehan and Sheehan (1982) have furnished a list of the major woody species of pine woods, maquis, and wood edges in the southern Argolid (Table 1). Unfortunately, the southern peninsula and the plant communities of the shore zone have received little or no detailed attention.

The natural vegetation, found on mountain slopes with shallow soil, along stream banks, on long-fallow fields, and along the seashore, is invariably strongly influenced by grazing. The gathering of herbs, greens, fruit and nuts (M. C. Forbes 1976b), and brush cutting for fuel and fodder exert further pressure, but data on their impact are lacking. The pine woods yield resin for wine, and thickets of wild olives are sometimes converted into olive groves.

Pine woodland (Plate 1a) dominated by *Pinus halepensis* is common in the area, although probably less so than in the recent past (Gavrielides 1976b). It covers many a ridge or hilltop, but it prefers soils derived from the ophiolite complex (see below) or young marls. A mature stand may be up to 8 m high, and its crown cover is complete so that pine seedlings cannot survive. Death from old age or by fire is the only means of regeneration for these woods (Rackham 1983). The undergrowth, usually much browsed, consists of Kermes oak, junipers, and thyme (*Coridothymus* sp.). A few pine woods appear to have been planted, and occasionally one sees pine seedlings taking hold in abandoned fields. It is uncertain whether pine woods are an original component of the natural plant cover in the area, or whether they are solely a product of human influence (M. C. Sheehan, personal communication). Pollen evidence indicates the presence of pine species since at least the Bronze Age (Sheehan 1979).

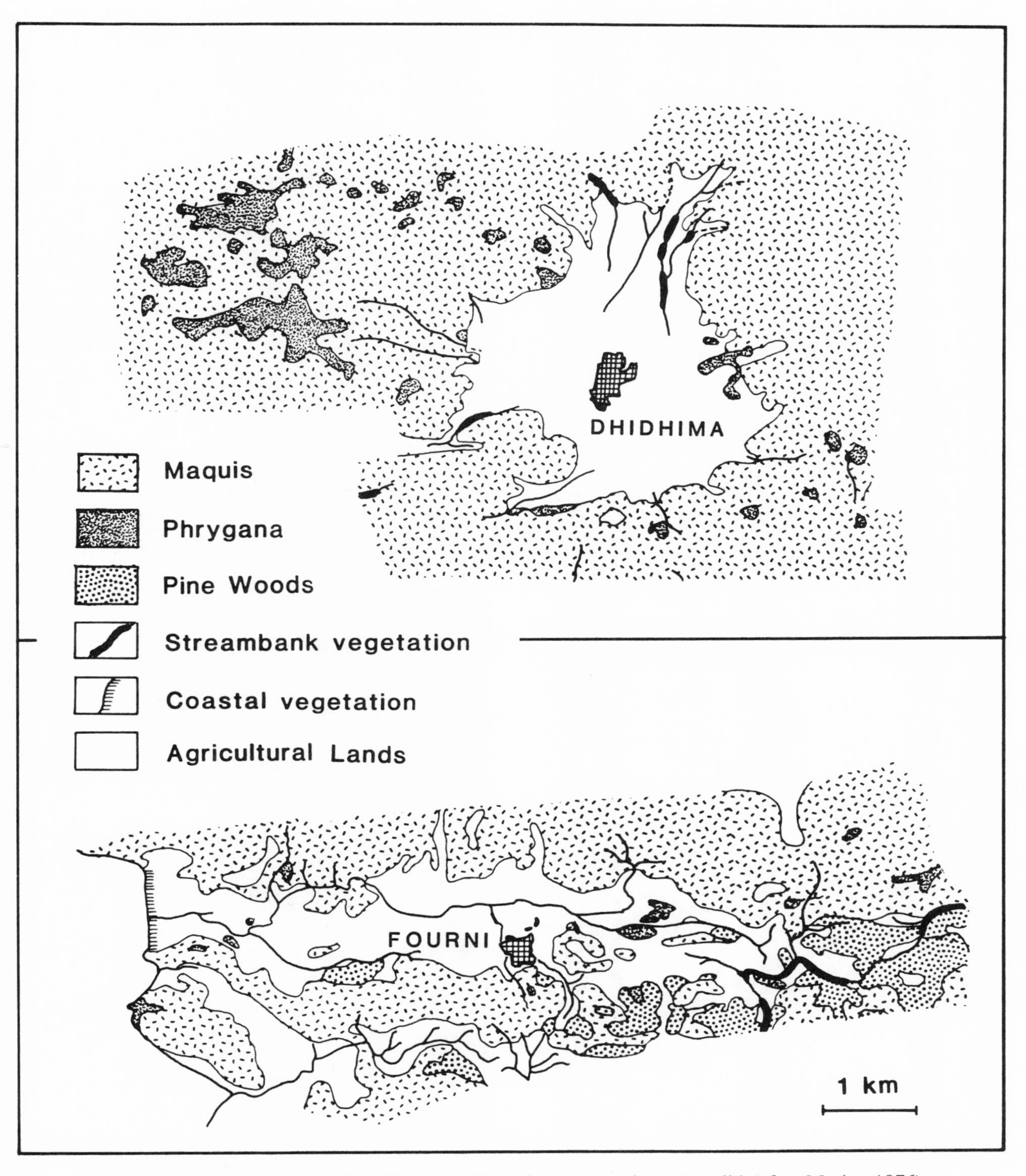

Figure 4. Vegetation in the Dhidhima and Fourni areas, southern Argolid (after Marius 1976).

TABLE 1

ASPECT-FORMING SPECIES OF THE NATURAL VEGETATION OF THE SOUTHERN ARGOLID
(After Marius (1976) and Sheehan and Sheehan (1982), with additions by van Andel)

Type	*Primary Aspect-forming Species*	*Other Aspect-forming Species*	*Accessory Species*
Pine forest	*Pinus halepensis*	*Juniperus phoenicea* *Coridothymus capitatus* *Genista acanthoclados* *Quercus coccifera* *Asparagus acutifolius*	*Erica verticillata* *Quercus ilex* *Myrtus communis*
Olive thicket	*Olea europaea*	??	??
Maquis	*Juniperus communis* *Quercus coccifera* *Pistacia lentiscus*	*Olea europaea* *Phillyrea media* *Coridothymus capitatus* on siliceous soils *Erica aborea* *Arbutus unedo* *Arbutus andrachne*	*Genista acanthoclados* *Pistacia terebinthus* *Sarcopoterium spinosum* *Calycotome villosa* *Phlomis fruticosa* *Origanum* sp.
Phrygana	*Cistus villosus, C. incanus* *C. salviaefolius* *Coridothymus capitatus*	*Ballota acetabulosa* *Poterium spinosum* *Salvia* sp.	*Centaurea minta* *Rosmarinus officinalis*
Wood hedges (W) and streams	*Quercus coccifera* *Pyrus amygdaliformis* (W) *Pistacia lentiscus* *Nerium oleander* *Vitex agnus-castus* *Platanus orientalis* (wet)	*Calycotome villosa* *Acer sempervirens* (wet) *Acer campestre* (wet) *Cotoneaster* sp. *Frangula alnus* (wet) *Spartium junceum*	*Asparagus acutifolius* *Smilax aspera* *Ruscus acutifolius*
Coastal flats	*Phragmites communis* *Vitex agnus-castus* *Salicornia* cf. *herbacea* *Arundo donax*	*Tamarix gallica* *Obione* sp. *Scirpus* sp. *Juncus* sp.	*Suaeda* sp. *Salsola kali* *Crithmum maritimum*

Marius (1976) noted the occurrence of a "forest" of wild olive (*Olea europaea*) on the slopes of the border ranges; patches of it can also be seen in the gorges of the upper Fourni valley. This is a special facies of the maquis, but little is known about its composition and habitat.

Maquis (Plate 2), the dominant natural vegetation type in the area, occurs in many forms, open or closed, tall or low, on limestones or on a siliceous substrate. Junipers tend to be the dominant species, with *Erica arborea* (heather) and *Arbutus* spp. (strawberry bush) on siliceous soils. Browsing limits the diversity of the maquis almost everywhere, and even *Quercus coccifera* (Kermes oak) and *Phillyrea media* (mock privet) can be suppressed. The wild olive is, as noted, locally common, and so is the carob tree (*Ceratonia siliqua*), but the latter may well have been derived from nearby trees cultivated mainly for fodder.

Phrygana (Plate 3a) is a community of low shrubs, resplendent with flowers in the spring. It is found on abandoned fields, stony slopes, and in patches within the maquis. Its low, dense, often roundish small shrubs are mainly thyme (*Coridothymus capitatus*), rock rose (*Cistus* sp.), and various kinds of broom (*Calycotome villosa, Genista acanthoclados*). The phrygana, though often regarded as an ultimate degradation stage, can also be a first step in a regeneration process towards maquis (Marius 1976; Shay and Shay 1978), and thence to oak woodland (Rackham 1983). Once the regeneration has reached medium height and dense cover, pine cannot take over.

Dense hedgerows dominated by small trees of pistachio (*Pistacia lentiscus*), wild pear (*Pyrus amygdaliformis*), and wild olive, as well as species of broom (e.g., *Spartium junceum*) form an almost impenetrable barrier along the edges of many agricultural terraces. Commonly, woodcutting and browsing have reduced it to a fringe of low, multi-stemmed shrubs, but where protected some of the components may reach tree height (Plate 1b).

Some of the same species, together with oleander (*Nerium oleander*), chaste tree (*Vitex agnus-casta*), and smoketree (*Cotinus coggygria*), all strung with thorny vines such as *Smilax aspera,* rim the dry stream beds. Where groundwater is shallow one finds the plane tree, *Platanus orientalis,* and at slightly higher elevations near permanent water small maples (*Acer campestre*), as on the Iliokastro plateau.

The valley floors and coastal plains have now been converted entirely to arable land, but in the past, besides oak woodland, wild grasses, herbs, and shrubs must have been common there on the flood plains, providing a useful array of root crops, greens, bulbs, cereals, and wild pulses (Shay and Shay 1978). This grassy cover may have extended up onto the more gentle lower slopes of many valleys. Of these communities nothing remains but scattered species often regarded as common weeds, except for what grows in the coastal marshes and salt flats at Thermisi, along Potokia and Kranidhi Bays and, no doubt, at one time along Kiladha Bay as well. These lands provide reeds and rushes (Plate 3b), the edible salt plant samphire (*Salicornia* cf. *herbacea*), and farther inland reeds (*Phragmites communis*), with patches of chaste tree and stands of tamarisk (*Tamarix gallica*). Freshwater marshes fed by springs once existed in several places along the shore, for example near Franchthi Cave, and still do elsewhere in the Peloponnese. Their vegetation, which includes freshwater rushes and reeds, willows, and grasses, is not well known.

GEOLOGICAL HISTORY

C. J. Vitaliano

Greece is a restless land plagued by numerous and often severe earthquakes, its shores sinking or rising almost measurably in some places, while some of its volcanoes have been active in the recent past. Among them are those of Methana and Aigina not far from the southern Argolid. Compared to such regions as the Gulf of Corinth, however, the southern peninsula itself is not very active seismically and does not appear to move up or down to any great degree (see, however, Flemming 1968).

The seismic and volcanic activity of Greece and the Aegean is due to their location at a point where the northward movement of Africa towards Europe interacts with a gradual westward shift of Asia Minor. This geological turbulence is not dissimilar to the endless migrations of peoples across this part of the world which brought upsets, burning, and the vanishing or appearance of entire civilizations.

The Franchthi-Ermioni region, currently an ''island'' of relative stability in the Aegean, is of interest to both archaeologists and geologists. Franchthi Cave, a limestone karst feature of great archaeological importance, is located in the western part of this region on a small bay opposite the fishing village of Kiladha. Geologically, the region is an elongate horst, an elevated crustal block bordered by faults on both long sides. These faults trend east-west and are of the reverse type. The northern fault separates pre-Cenozoic rocks of the so-called Thessalian Crystalline Complex (Dürr et al. 1978) in the north from later Mesozoic and early Cenozoic deposits in the south. The southern fault is between late Mesozoic and early Cenozoic deposits in the north and a volcanically derived Mesozoic complex with Late Cenozoic sedimentary rocks in the south.

The rich artifact record unearthed in Franchthi Cave raises many geological questions which, in addition to the history of the landscape, include the sources of raw materials for the manufacture of stone tools. In late July and early August, 1979, a 12-day field study of the bedrock in the region was undertaken in order to determine the extent to which these raw materials may have been locally derived. The map and cross sections in Jacobsen and Farrand (1987: Plate 1) are the result. The map covers 125 km^2 of the area between Dhidhima in the north and Kranidhi in the south. In this section, the distribution and nature of the geologic units which underlie the region and outline its principal structural features will be discussed. The youngest deposits, those of the late Quaternary, will be dealt with in the next chapter. An attempt to answer questions concerning the source of at least some of the raw materials for the artifacts will be made in the closing section of this chapter.

Formations and Stratigraphy

The Ermioni region is underlain by five major geological formations (Jacobsen and Farrand 1987: Plate 1): an older, thick-bedded, massive weathering, sparsely fossiliferous, and slightly cherty dolomitic limestone of Middle Jurassic age, a series of serpentinized igneous rocks and intercalated thin-bedded limestones of Late Jurassic to Early Cretaceous age, a series of thin-bedded limestones and marls with associated thick-bedded, karstified shallow-water limestones of Early Cretaceous to Paleocene age, thick-bedded turbidites, sandy limestones and calcareous marls of Paleocene to Eocene age, and a thick sequence of limestone-cobble and gravel conglomerates and marls of Neogene age. All of these units are in places covered with hillwash and locally derived erosion products. Finally, alluvium fills present stream channels and covers the small coastal plains.

The small area of study, the sparse fossil content of its rocks, and the general uncertainty concerning the position of the units within the stratigraphic framework of the Argolid as a whole renders assignment of the units to their specific levels in the geologic column difficult. However, on the basis of the fossils present in the basin and of regional relationships of the strata, together with fossil evidence from other parts of the Argolid, Bachmann and Risch (1979) erected a stratigraphy for the entire Argolid Peninsula. The ages assigned below to the various stratigraphic units are based largely on their findings.

Pantokrator Limestone (Middle Jurassic). The Pantokrator limestone crops out over large areas in the northern part of the area. It makes up the high mountain range extending from Megalovouni west to the sea near Vourlia bay. It is considered part of the Pelagonian (*sensu stricto*) nappe (Dürr et al. 1978), correlatable with the massive limestones of the same name in continental Greece (Jacobshagen et al. 1978). The true thickness of the formation in the area is unknown, but at least 500 m are exposed near Dhidhima.

The Pantokrator is composed of a dark gray, massive-weathering, fine-grained dolomitic limestone. The presence of coralline fossils suggests reef-type deposition. Locally, the large clam-like fossil *Megalodontides durga* (?) and oolites are present. In places, small patches of recrystallized limestone fill gashes in the body of the rock. To the west of Dhidhima, along the northeastern shore of the Franchthi embayment, an extensive outcropping of limestone breccia is present at the foot of the steep southern slope of the Dhidhima Range. Outcrops of similar breccias are found at the base of the limestone in several narrow valleys cut into the Dhidhima Range, in the upper Fourni valley and farther east.

Ophiolite ''Nappe'' Series (Middle Jurassic). The Pantokrator limestone is overlain by a series of altered mafic rocks and associated clastic and volcaniclastic sediments referred to either as the Diabase-Hornstein-Tuffite Series or as the Ophiolite ''Nappe'' (Bachmann and Risch 1979). Within the Franchthi-Ermioni region the series consists of serpentine and basalt (including pillow lavas) with related tuffaceous sediments, together with bedded, black and chocolate-brown radiolarian cherts and purplish, thin-bedded limestones. As a whole, the series is typical of an oceanic sequence. Its total thickness in the area is uncertain, due to the intervention of the major reverse fault which forms the northern boundary of the basin.

The serpentine, a dark greenish or dark brown to black, massive weathering, highly fractured rock, underlies most of the Fourni valley, making up all of its lower elevations as well. A narrow ridge of carbonate sediments which extends discontinuously along the entire length of the Fourni valley divides the serpentine into two bodies. Pillow lavas of marine origin and basaltic composition are associated with the serpentinite. They crop out along the southwest shore of the Franchthi embayment, just west of Kiladha, as bulbous, pillow-like masses of highly altered basalt. Alteration of the mafic parent rock is so widespread that only a single sample of unaltered basalt was encountered northwest of Kiladha. Dürr et al. (1978) referred to these rocks as the ophiolite "nappe."

The serpentinite characteristically weathers very rapidly, but the smooth, rounded, low terrain it underlies easily betrays its presence. Except for the single sample noted above, no unaltered rock was encountered over most of the study area. Nevertheless, examination under the petrographic microscope revealed that remnants of some of the minerals from which it was derived are still present. Olivine and clinopyroxene, probably augite, were host minerals that have been positively identified. In addition, a spinel mineral, either magnetite or chromite, and sparse flakes of bleached mica were encountered. All suggest that the host rock was diabase or peridotite. Extensive fracturing of the rock and subsequent hydrothermal alteration have replaced most of the original minerals with serpentine (essentially a hydrous magnesium silicate mineral). Later fractures within the serpentinite proper have been filled with an asbestiform variety of serpentine (chrysotile) and/or carbonate.

The basalt, including the pillow lava, is a very fine-grained, somewhat fractured rock with sparse white and greenish coating. The fractures are filled with carbonate. The quenched nature of both basalt and pillow lava becomes evident when examined under the microscope. In thin section, clusters of needle-like skeleton crystals are embedded in a glassy ground mass which indicates rapid chilling of molten lava upon entering the water.

The Franchthi Limestone (Early Cretaceous/Cenomanian). In the Franchthi-Ermioni region prominent features are two ranges of hills trending east-west. One extends from the Franchthi headland eastward, the other runs parallel to it north of the Kranidhi-Ermioni road from Profitis Ilias to just west of Ermioni. Both ranges are made of limestone and marl divisible into two units. The lower unit is composed of a mixture of limestone and marl containing occasional fragments of the underlying ophiolite "nappe." It is thin-bedded and, in this area, 15-20 m thick at most. The upper unit is a gray-black, massively bedded limestone with occasional thin stringers of blue chert (Van Horn 1973). The erosional surface of this limestone contains numerous very fine protuberances of greater resistance, probably composed of a mixture of clay and silica. In the Franchthi-Ermioni region this unit is approximately 100-150 m thick.

Under the microscope, the rock is technically a fossiliferous micropelite, that is, an exceedingly fine-grained mass of calcium carbonate mud enclosing fragments of limestone and fossil shells, and pellet-like aggregates of calcium carbonate. The massive nature of this rock, a prominent field characteristic, carries through to the thin sections where the mud and clay-sized fragments are packed together without evidence of layering. However, the rock is considerably fractured and the fractures are filled with sparite, a coarse-grained recrystallized calcium carbonate.

Franchthi Cave, a large karst feature on the north shore of Kiladha Bay, is located in this massive unit. Accordingly, the entire sequence has been designated informally as the Franchthi limestone. Bachmann and Risch (1979) have assigned an early Cretaceous (Cenomanian) age to the formation.

Limestone-Marl Sequence (Late Cretaceous to Early Paleocene). The low chain of hills just south of the Kranidhi-Ermioni road extending from Kranidhi eastward to Ermioni consists of 200-300 m of thin-bedded to platy limestones, marls and siliceous mudstones of Late Cretaceous to Early Paleocene age (Bachmann and Risch 1979). The lower part of this sequence, exposed in the base of the north slope near Ermioni, is composed of thin-bedded, dark gray, siliceous mudstone and fine-grained limestone. Under the microscope, the limestone is technically a biomicrite consisting of minute fossils (Radiolaria and Foraminifera) embedded in a very fine-grained carbonate matrix. The upper part of the sequence is a reddish to gray to greenish gray calcareous marl with associated limestone. The bedding in both parts of the sequence is tightly folded, and the fold axes are disoriented to a high degree. The nature of the folds suggests either soft-sediment deformation just after deposition, or postdepositional tectonic readjustment, as indicated by the disorientation of the fold axes.

Flysch (Late Paleocene to Eocene). The central part of the Franchthi-Ermioni region is underlain by a sequence of gray to pinkish gray to greenish brown, poorly fossiliferous, thin-bedded, graded deposits composed chiefly of marls, sandy and calcareous shales, and mudstones which are rhythmically interbedded with coarser-grained sediments. The entire association, designated the Flysch, has been assigned a Late Paleocene to Eocene age; farther east it makes up virtually the entire Adheres Range.

Flysch outcrops are rare in the Franchthi-Ermioni region due to the extensive cover by alluvium, itself derived largely from the Flysch. The Flysch probably underlies much of the lower parts of the region, a supposition supported by the scattered outcrops of this formation in the Loutro and Ermioni valleys as well as in the valley between the Profitis Ilias-Ay. Apostoli range of hills and the Kranidhi-Ermioni range. Flysch outcrops were also noted along the perimeter of Koronis Island, but they could not be examined further because the island is privately owned.

Based on the size of the individual grains (0.0625-2.0 mm), some of the Flysch rocks can be classified as sandstone (Nockolds et al. 1978). They include quartzo-feldspathic arenite cemented with quartz, and calcarenite (composed of more than 50% detrital carbonate in sand-sized grains). The arenites are rhythmically interbedded with marls and calcareous shales which contain a few fossil fragments and sparse grains of quartz. The thickness of the Flysch in the Franchthi-Ermioni region is unknown.

Later Cenozoic Deposits. A series of chalky limestones, sandstones, conglomerates, and clays of later Cenozoic age, probably late Miocene to Pliocene (Süsskoch 1967; Bachmann and Risch 1979) underlies most of the southern and southwestern part of the southern Argolid. These rocks were deposited unconformably upon the Flysch and the ophiolitic sequence. Forney (1971) divided them into a series of northeasterly-dipping chalky limestones, clays and fine-grained conglomerates called informally the Dhouroufi formation, and a series of southwesterly-dipping limestones, sandstones, limestone conglomerates, and clays called informally the Metokhi formation. The combined thickness of these two units does not exceed 350-500 m.

The Dhouroufi formation is represented by scattered outcrops in the area west of Kranidhi and south of Kiladha. The Metokhi formation, especially the limestone conglomerate members, forms the prominent east-west escarpment on which the town of Kranidhi is situated. The conglomerate consists of well-rounded cobbles and boulders of limestone set in a limestone matrix. The finer sediments of this unit are calcarenites (a calcareous detrital sand with 0.0625-2.0 mm grain size), limestones, and clays.

Breccia (Quaternary?). A thick limestone breccia crops out along the base of the massive limestone in parts of the Franchthi-Ermioni basin. The major deposit is found along the south slope of the Dhidhima Range between Akra Ay. Nikolaos and the Ay. Ioannis Prodhromos valley. Scattered outcrops also occur near Omeni, 3 km northwest of Ermioni, at the foot of the northeastern slope of Mt. Asprovouni, 3.5 km northeast of Kranidhi, and in the upper reaches of the twin valleys at Karakia, 1.5 km northeast of Fourni.

The breccia, in all instances, is composed of fragments up to one meter in diameter and derived from the massive limestones. In the Omeni and Asprovouni localities the source appears to have been the Franchthi limestone, shed undoubtedly from the high hills in this area. The outcroppings along the front of the Dhidhima Range and in the valleys at Karakia may have been derived from the Pantokrator Limestone, but here the origin is more obscure. The outcrops may represent a form of solution breccia, a cemented talus deposit, or even a cemented fault breccia. The large amount of red matrix and the angularity and dimensions of the fragments deny a solution breccia origin and support that of a cemented talus. However, an origin as a fault breccia cannot be dismissed without further study. Movement along the thrust fault that marks the northern boundary of the Franchthi-Ermioni region must also be considered as a possible mechanism.

Age and Correlation of the Formations

In view of the scarcity and unidentifiable nature of the fossil remains noted during the course of field mapping, what little information has been presented regarding the age and correlation of the various formations above has been gleaned from the recent literature, especially from Bannert and Bender (1968), Süsskoch (1967), Forney (1971), Dürr et al. (1978), Jacobshagen et al. (1978), and Bachmann and Risch (1976, 1978, 1979).

The Franchthi-Ermioni region of the Argolid Peninsula is part of the Attic-Cycladic crystalline complex of the Aegean orogenic arc (Dürr et al. 1978). It is underlain by carbonate, siliceous, and volcanogenic sediments and altered igneous extrusive rocks ranging in age from middle Mesozoic to Holocene.

The oldest formation exposed in the mapped area is the massive limestone which crops out along the northern margin of the map. On the basis of studies outside the map area, this formation has been designated the Pantokrator Limestone (Jacobshagen 1972; Dürr et al. 1978; Bachmann and Risch 1976, 1978, 1979) of Middle Jurassic age and has been correlated with similar rocks which crop out over large areas of the Peloponnese and the Greek mainland.

The complex of serpentinites and remnants of the rocks from which it was derived (peridotites, diabases, and pillow lavas, volcaniclastic and carbonate sediments named the ophiolite "nappe") has been identified as Middle Jurassic to Early Cretaceous in age (Bachmann and Risch 1978, 1979). The serpentinite has been assigned to the upper part of this interval and explained as having been derived from a large, tectonically emplaced sheet of oceanic crust composed of mafic igneous rocks. According to Bachmann and Risch (1979), the earlier Pantokrator Limestone was uplifted and karstified at this time.

After the uplift and karstification of the Pantokrator Limestone, Early Cretaceous time saw the deposition of a series of pelagic carbonates in the area (Bachmann and Risch 1978, 1979). The massive unit of the formation in which Franchthi Cave was formed gave it the informal name of Franchthi limestone.

Subsidence continued into Late Cretaceous time when the reddish to whitish cherty limestones and red marls were deposited. These either underwent soft sediment deformation or were deformed afterwards into a series of randomly oriented minor folds. The upper units of this formation constitute part of a transition zone into an early flysch-type deposit of Paleocene age (Bachmann and Risch 1979). Quite possibly, the eroded, reddish, moderately cherty limestones which crop out at the base of Khonia and at Akra Kokkinos were laid down at this time. Subsequently, the main Flysch deposits in the Franchthi-Ermioni region were laid down in Eocene time.

During Miocene time the finer-grained sediments and cobble conglomerates of the Dhouroufi and Metokhi formations are thought to have been deposited (Forney 1971). These rocks crop out nearly everywhere in the southern part of the southern Argolid, and their conglomerates form prominent escarpments, for example, the one trending westward from Kranidhi.

Structural Geology

Structurally, the Franchthi-Ermioni region is a horst (Bachmann and Risch 1979), an uplifted block of strata bound on the long sides by faults. The cause of this structural pattern lies in its position in the Argolid and the Peloponnese within the Mediterranean structural complex. Tectonic readjustment of the southern part of the European and the northern part of the African continents in response to plate movements has deformed the major rock units in the region. Deformational effects may be represented by the tightly folded rocks that form the Rakhes-Profitis Ilias ridge between Kranidhi and Ermioni. The near vertical and commonly overturned attitude of the fold axes in the thin-bedded sediments and the closeness of the folds suggest severe deformation, although soft-sediment deformation cannot be categorically dismissed.

The outcrop pattern of the rocks is generally simple: all units trend east-west. Contact relationships between the thin-bedded Flysch units and the ophiolite ''nappe'' are obscured almost everywhere by the presence of the massive limestones (either the Pantokrator or the Franchthi) or by their debris. In the relatively low ridge that bisects the Fourni valley, however, the relationship between the serpentinite and the sequence of thin-bedded limestones which form the backbone of the ridge is observable. The limestones crop out discontinuously throughout the length of the valley from Ay. Ioannis on the Franchthi embayment in the west to Ay. Ioannis near Iliokastro in the east. All contacts between the serpentinite unit and the thin-bedded limestone that were examined appeared to be devoid of the effects of thermal metamorphism, suggesting that the serpentinite or, more likely, the rocks from which it was formed had been emplaced and crystallized prior to the deposition of the sediments.

The region is bounded on the north and south by east-west trending reverse faults (Jacobsen and Farrand 1987: Plate 1). The northern fault dips to the south and extends from the north side of Koronis Island to beyond the eastern boundary of the map. Its hanging wall over much of its extent is the Franchthi limestone. The footwall is composed of serpentinized mafic igneous extrusive rocks of the ophiolite ''nappe.''

The southern fault dips steeply to the north and extends from near Kranidhi on the west eastward along the valley to the eastern boundary of the map (Jacobsen and Farrand 1987: Plate 1) near the northwest corner of Ermioni. The hanging wall is composed of Flysch sediments, rocks of the Franchthi limestone, and valley alluvium. The footwall rock outcrops are largely obscured by alluvium. Where outcroppings are observed, they consist of the same rocks that make up the Kostelaneïka ridge.

Several minor faults, both normal and reverse, occur within the area. One prominent reverse fault crops out on the western slope of Profitis Ilias, trending northwest and dipping to the west. Its displacement is small but noticeable, and it involves only rocks of the Franchthi limestone. The remaining faults are mostly normal and of small displacement. They are located commonly near contacts between formations. Undoubtedly, similar structural features are present in the Flysch and the volcanically derived portion of the "ophiolite" sequence, but their traces are often difficult or impossible to recognize. The northwest-trending fault 1.5 km west of Mavrovouni and the northwest-trending one near Ay. Ioannis are exceptions.

WATER AND OTHER RESOURCES

Tj. H. van Andel and C. J. Vitaliano

Water is in short supply in the southern Argolid owing to the very low annual rainfall, the long dry season, and the underground drainage that is normal for a land with so much limestone. The streams are mostly dry, flowing only after a major rain and then not every year. Only on the Iliokastro plateau have we found streams running as late as September, due to the presence of large springs at the foot of the Megalovouni massif. Farther east, small spring-fed streams are found in some of the upper valleys of the Adheres range, but few are reliably perennial. The presence of several Turkish watermills in the upper Ermioni river valley and in the Adheres range on now generally dry streams suggests that in the eighteenth and early nineteenth century runoff, and therefore precipitation, may have been somewhat higher.

We have no runoff estimates for the area, but informants say that surface flow takes place during exceptionally heavy rains and after the soil has become saturated in autumn and early winter. Even the larger streams cease to flow after a few days or at most a few weeks. Drainage basins are small (Figure 5). The largest is that of Ermioni (70 km^2), and the majority measure less than 10 km^2 (Pope and van Andel 1984: Fig. 1).

In many streams subsurface flow in the gravel and sand bed continues long after it has ceased at the surface, and as late as July we have seen water collect in shallow holes. Springs also exist in some stream beds where the permeable fill abuts against a rock ledge downstream. Such places are a valuable resource in semiarid or arid regions, a fact well known to the shepherds and farmers of the southern Argolid today.

Most of the precipitation, falling on rather bare limestone slopes, enters an underground drainage system that may have had its origins at the time of the middle Cenozoic uplift of the region. The seepage collects above an impermeable base of Flysch or ophiolite, and where the contact crops out, rows of springs form, for example along both sides of the Fourni valley. Most of these springs are now dry because innumerable wells have drawn down the level of the groundwater, but their former presence is betrayed by deposits of travertine. North of Iliokastro and on both sides of the Adheres range, more copious springs still exist because there is less agriculture and hence there are fewer wells in that area. They bear witness that the region was not always as devoid of usable water supplies as it is now.

The contact between the limestone and the ophiolite or Flysch also crops out below sea level, and submarine freshwater springs are quite common in the area (Figure 5; van Andel and Lianos 1983). The spring water is easily sensed by swimmers and divers because it is much colder and less buoyant, and many shallow submarine springs near the shore have so been identified. Among them are those at the Franchthi shore which provided a ready water supply for wet-sieving during the excavation. Most of these springs have only recently become submerged by the rising sea. Until then, they fed small but perennially green meadows and shrubbery on loamy coastal terraces such as the one once existing below the entrance of Franchthi Cave (see Chapter Three, Figure 21).

Submarine springs can also be detected by marine seismic reflection profiling, because the boundary between the warmer saline and cold fresh waters is a good reflector (van Andel and Lianos 1983: Fig. 7). No fewer than seven such springs have been found in depths ranging from −20 m to −185 m along the relatively few geophysical traverses around the southern Argolid, indicating that they are common indeed (Figure 5).

It has been suggested that at lower stands of the sea submarine springs such as those below Franchthi Cave would have drained underground, because they would owe their existence to an impermeable barrier of marine sediments resting against the limestone. This seems unlikely. The recent nearshore deposits of the area are thin and far from impermeable, and most of the springs issue from a contact well above the surface of the old Pleistocene coastal plain.

At present the need for water in the area is satisfied virtually entirely by wells, an ancient practice here. The oldest ones still in use date back to Classical times but it seems likely that the digging of wells became necessary much earlier, during the later Early or Middle Bronze Age, when the southernmost peninsula, where springs are rare, became densely settled. An overview of the use of groundwater and the history of wells in the area can be found in Harper (1976). At the present time, large-scale withdrawal of water from wells in the coastal plains has resulted in considerable saltwater invasion, as it may have done at times in the past.

Therefore, barren as the Franchthi region may appear at first sight, not so long ago it possessed modest resources of water, acceptable soils (Chapter Two), and usable vegetation and associated wildlife, as well as some marine resources (see Chapter Three). What it has little of are exploitable rocks and minerals. Small mineralized zones with iron and manganese

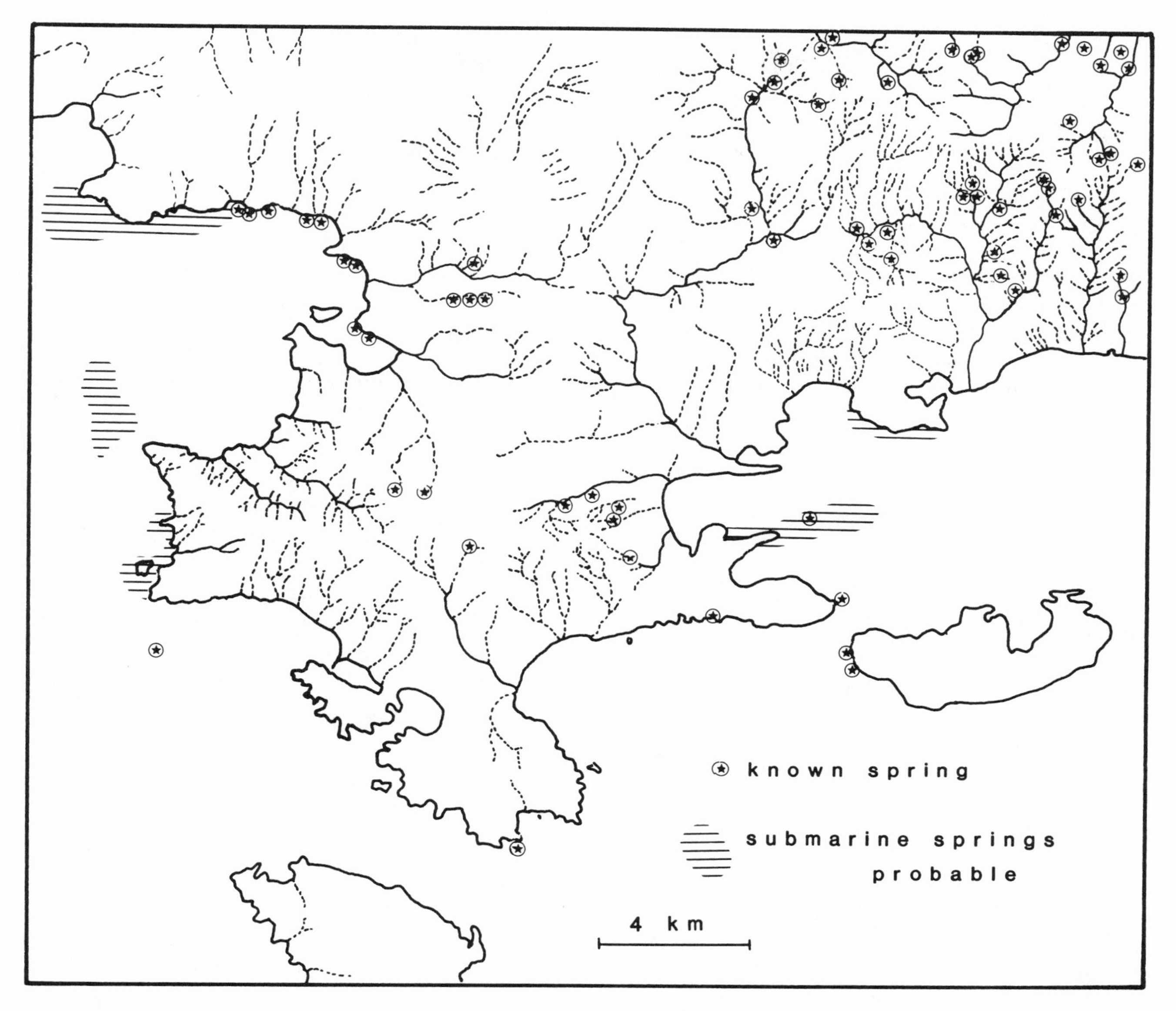

Figure 5. Streams and springs of the southern Argolid. Major streams are shown with solid lines, dry washes dashed. Drainage and springs on land have been taken from the 1:50,000 Greek Army Geographic Division maps 50,000 (1964 edition), sheets Idhra and Spetsai, which represent the situation on or before 1940. Additions from Harper (1976) and our own observations. Submarine springs from various coastal explorations and from van Andel and Lianos (1983).

pyrites occurring in the Adheres range between Iliokastro and Thermisi have been exploited intermittently and on a small scale for the past century (Aranitis 1963; Aronis 1938, 1951; Moussoulos 1958; Voreadis 1958). The same ore bodies also may contain up to two percent copper, but we have no direct evidence that this metal was mined in prehistoric or historic times. The semiprecious copper minerals, such as malachite, that were used for decoration by the inhabitants of the Franchthi Cave, however, may well have been obtained from outcrops of these ores, just a few hours walk away.

More mundane materials of local origin have been used for many millennia. The cherts from the ophiolite complex, especially the chocolate-brown kind, have been used as tool materials since the Middle Paleolithic (Pope et al. 1984), and at least one site of probable prehistoric age is known where this material was quarried (Jameson et al. forthcoming: site F25). The lower Flysch unit contains light-colored cherts which may have been the raw material for other tools. Furthermore, Van Horn (1973; Jacobsen and Van Horn 1974) refers to a "blue" flint which occurs in "tabular form, i.e. in long, thin strata sandwiched between layers of chalky limestone" (Jacobsen and Van Horn 1974: 306). This formation crops out on top of a small hill (Paliokastro) in the Fourni valley ca. 2.5 km northeast of the Franchthi cave. The chert is similar to light-colored cherts found elsewhere in the massive limestone. Fragments of blue chert are also found in the beach gravel at the mouth of the Fourni river, presumably derived from the Paliokastro source.

In general, the cherts tend to fragment into very small pieces suitable only for microscale implements (Jacobsen and Van Horn 1974), due to the intense deformation of the ophiolites and the fracturing of the cherts of the Flysch perpendicular to the bedding of the limestones. Quarrying these cherts for the making of larger tools must have been an unrewarding practice, which potentially encouraged the manufacture of the smallest possible implements. Although such small implements have been found, they are by no means the only ones. Larger, less flawed pieces of chocolate-brown or light-colored chert must have been used for the remainder. They probably came from stream cobbles, still occasionally seen but now rare.

Artifacts of fair size made from serpentine, diabase, and basalt have also been found in the cave. The appropriate rocks occur in the ophiolite complex and in volcanic bodies in the Dhiskouria hills southwest of Ermioni and northwest of the Franchthi embayment at Vourlia. Everywhere, however, these rocks are deeply weathered, and the acquisition of large pieces of fresh material from outcrops would have been very difficult. Once again, stream cobbles, concentrated in the float by natural processes which eliminated the fractured or weathered material, are a likely source. At the present time, sound cobbles are rare, but they may have been more abundant in the past, having become depleted by human exploitation (Perlès 1987). Even after exploitation ceased, erosion and stream transport would require many thousands of years to replenish the supply.

Grindstones and millstones may on occasion have been made from local cobbles as well. The Flysch contains suitable sandstones resembling those utilized, but no local source has thus far been demonstrated for the Franchthi finds, and importation from elsewhere remains a possibility, perhaps when suitable stream cobbles became scarce. With the passing of time, grindstones were increasingly made from a porphyritic andesite imported from Aigina (Runnels 1981, 1985). Both the Dhiskouria hills and the Vourlia area have porphyrite stocks resembling those of Aigina and might have been used for the same purpose, but no attempt has yet been made to identify them among the assemblage of artifacts.

Finally, no source of marble has been found in the region, although small veins of coarsely crystalline calcite in the Franchthi limestone might have been used for smaller objects. For others, importation from distant but more abundant sources such as exist near Navplion appears probable.

CHAPTER TWO
Soils and Alluvium

EROSION, ALLUVIATION, AND SOIL FORMATION

Between 1979 and 1982, a detailed study of late Quaternary soils and alluvium was carried out in the southern Argolid in the context of the Stanford Archaeological Survey, taking as its starting point the work of Drost (1974). The details of this field study and its methods have been fully discussed in Pope and van Andel (1984) and need not be repeated here. During this survey, all of the many drainages in the southern Argolid (Figure 5) were mapped on the ground after a preliminary analysis of aerial photographs. A soil and alluvium map of the Franchthi area is presented in Figure 6; those of other parts of the region can be found in Jameson et al. (forthcoming).

The streams in the southern Argolid have laid down a discontinuous alluvium locally grading upward into slope wash (colluvium) deposits. Deposition and stream incision in the valleys, and erosion and soil formation on the slopes have alternated, leaving a complex late Quaternary record that has been deciphered mainly with the aid of paleosols, ancient soil horizons that serve as stratigraphic markers. In addition, geomorphic features such as terraces and fan lobes, and the nature of the alluvial deposits themselves (their facies) were used. The deposits were dated with a variety of archaeological and isotopic methods, with full recognition of the limits of both. Sites superimposed on deposits provided several dates *ante quem,* whereas artifacts contained within them yielded *post quem* dates. Isotopic dates were obtained on pedogenic carbonate nodules and crusts with the uranium disequilibrium method (Ku and Liang 1983). The chronological data are few (Pope and van Andel 1984: Table 3), but the paleosols themselves, besides being useful as regional stratigraphic markers, are also indicators of relative age. Together, they have established a reasonable chronological framework, have set useful limits for the onset and termination of alluvial events, and have confirmed the validity of correlations between various drainage areas.

The alluvial and colluvial deposits of the southern Argolid exhibit five different depositional types or facies (Figure 7). The first group is formed by the *debris flows* which are deposited rapidly, often catastrophically, from a slurry of clay, silt, and sand containing gravel, cobbles, and boulders. Its deposits are easily recognized by the fact that the chaotically dispersed coarse components appear to float in a matrix of fine sand, silt, and clay. There is no sorting, but in general the size of the coarse components diminishes upward in any single unit.

Stream-flood deposits form in braided stream channels and on bars. They consist of discontinuously stratified sequences of poorly to moderately sorted sand, gravel, or cobbles. The coarser components are usually subrounded. The beds contain much fine sand and silt, but rather than supporting the coarser components as in debris flows, this finer fraction merely fills interstices or forms separate beds. The distinction between debris flows and stream-flood deposits is therefore usually straightforward.

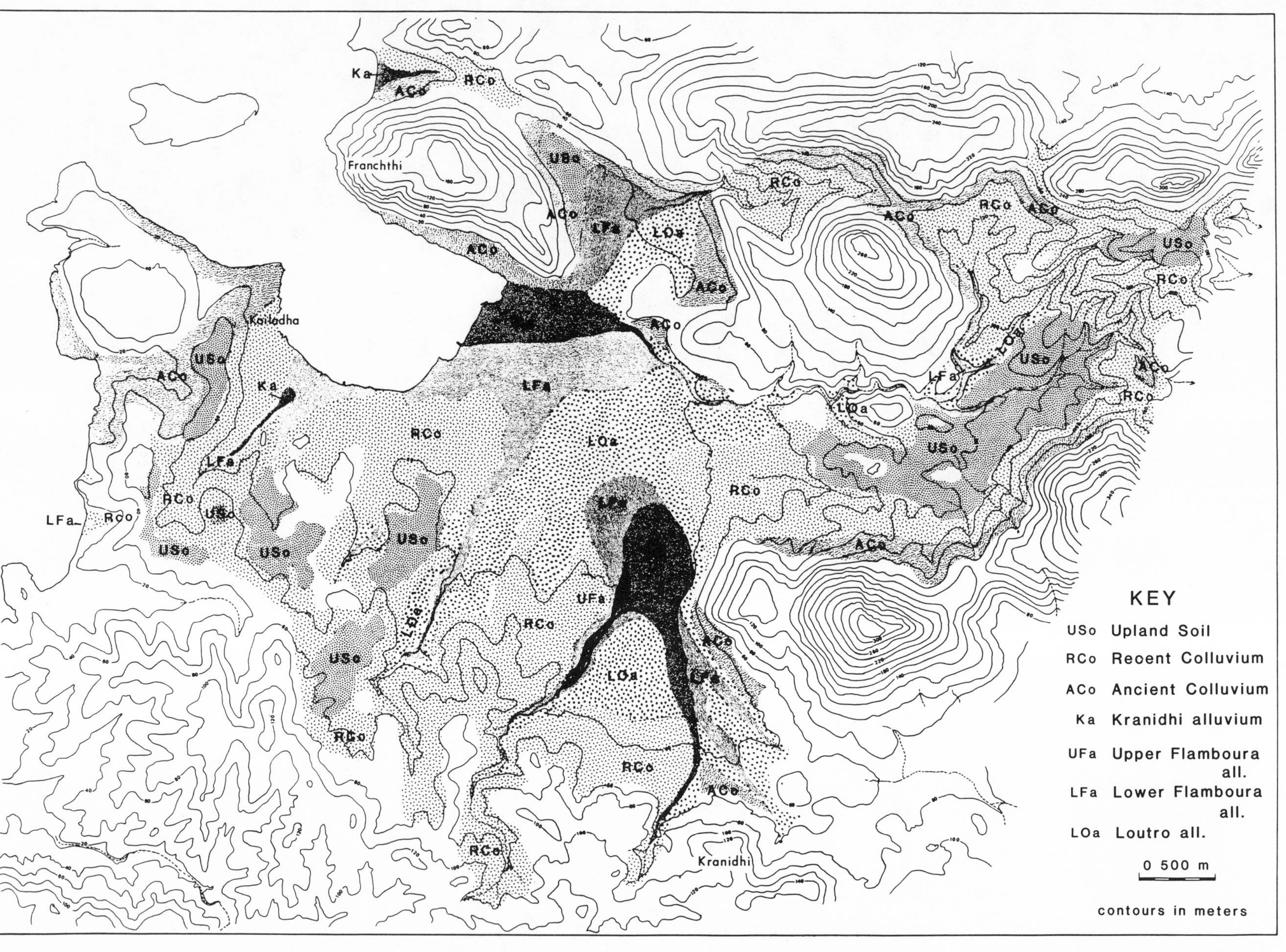

Figure 6. Late Quaternary alluvium and soils of the Franchthi area. Contour interval 20 m; contours from Greek topographic maps, scale 1:5,000. For complete stratigraphic sequence, see text and Figure 8. After Jameson et al., forthcoming:Chapter 3.2).

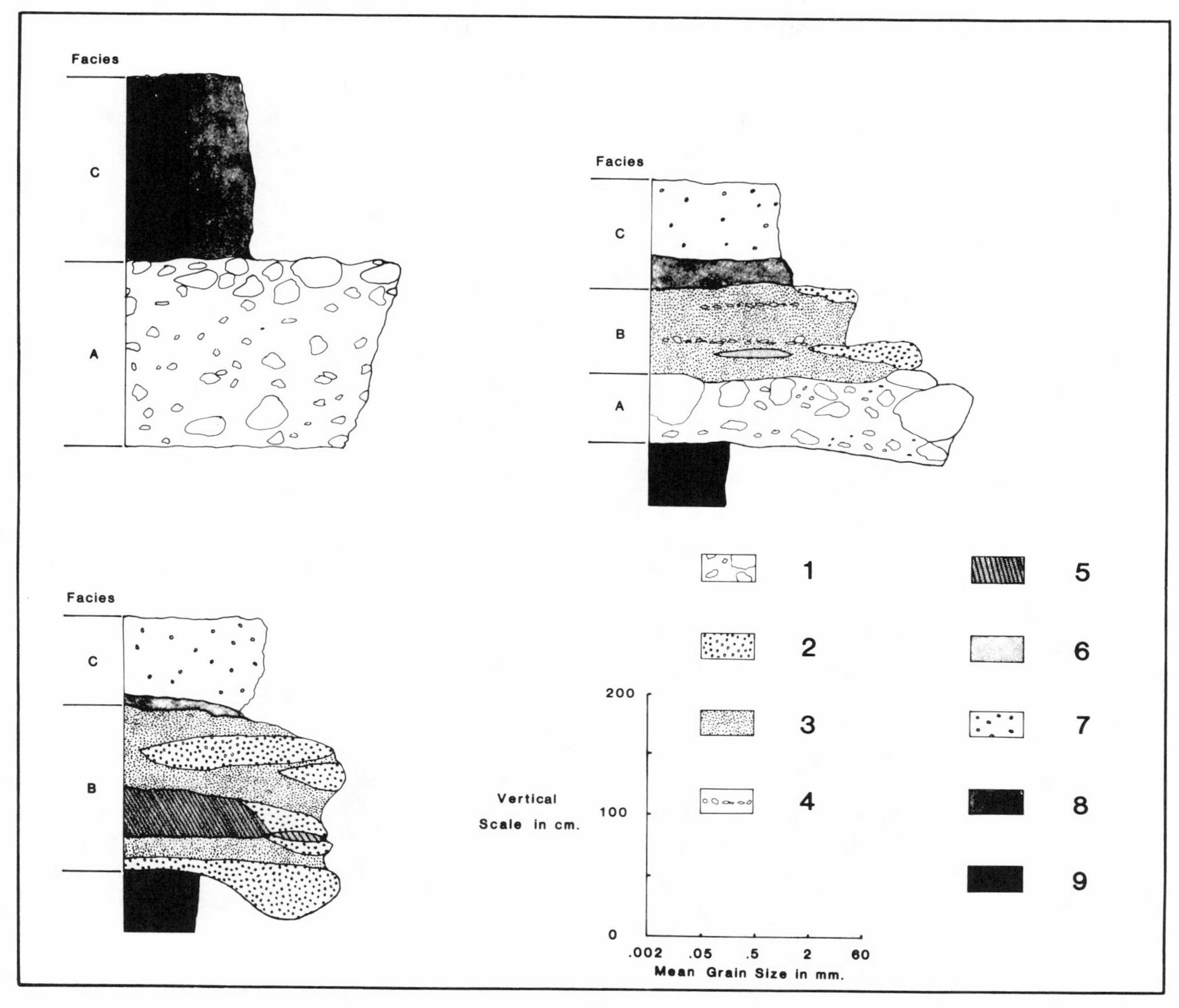

Figure 7. Facies sequences of alluvial deposits in the southern Argolid. Facies *A*: debris-flow deposits; *B*: stream-flood deposits; *C*: overbank loam. *Upper left: A-C* sequence in upper Flamboura alluvium; *upper right: A-B-C* sequence in Pikrodhafni alluvium; *lower left: B-C* sequence, lower Flamboura alluvium. *Key: 1,* large cobbles and boulders supported by a clay and silt matrix, crude inverse graded bedding: debris-flow deposit; *2,* lenticular, imbricated coarse gravel and cobbles with little or no fine matrix: stream-flood deposit; *3,* sand, gravel, and cobbles, interstices filled with clay; *4,* stringers of single pebbles or cobbles; *5,* cross-bedded sands and gravels from a channel bar; *6,* lenses of sand or fine gravel; *7,* sandy loam, little or no bedding: overbank loam; *8,* buried soil horizon, often truncated; *9,* pedogenic carbonate horizon.

The most extensive deposits, the *overbank loams,* consist of sandy loam with scattered pebbles and thin stringers of fine gravel. They form during peak floods and sometimes as a result of slope wash along the edges of valleys. The upper part of an overbank loam deposit tends to be free of coarse material, and stratification is usually indistinct or absent. This is partly due to the slow and intermittent deposition on a vegetated surface and partly to the mixing of the soil by roots and burrowing animals. Overbank loams form the surface of most valley floors and all coastal plains.

The overbank loams can be up to 5 m thick, especially in the lower valleys and on the coastal plains. The thickness of the stream-flood and debris-flow deposits varies from a few tens of centimeters to a few meters, and their lateral extent is measured in tens to hundreds of meters. A few debris flows can be traced well enough to show that a single event blanketed an entire valley bottom with a sheet several meters thick. Debris flows have not been observed in the process of formation in the area, but stream-flood and overbank-loam deposition are active at the present time.

Colluvium is a poorly sorted loam with variable amounts of mostly angular pebbles and cobbles. It is not always easily distinguished from overbank loam except by its position on a slope or where it contains the ''stone lines'' characteristic of this facies. Much rarer is a colluvium of angular cobbles and boulders within a matrix of silt which changes upward into lenses of pebble-sized, sharp-edged rock chips of fairly uniform size with few fines. The chips resemble frost-spall material. Pope and van Andel (1984) concluded that these deposits are a local form of the *grèzes litées* (Butzer 1964; Washburn 1980) which imply a cold climate with periglacial conditions prevailing at the elevations from which the chips were derived. The lenses of this facies tend to be 50-100 cm thick, and the chips measure a few centimeters.

The five facies compose themselves into orderly sequences that make up the seven late Quaternary stratigraphic units recognized in the area. A complete sequence would, from bottom to top, consist of debris flows, stream-flood facies, and overbank loams, but abbreviated ones occur (Figure 8). Virtually always and everywhere, however, the sequences terminate with an overbank loam in erosional contact with the overlying unit. The proportions of the several facies vary considerably from one stratigraphic unit to the next, although in general the bulk of each consists of overbank loam.

Each of the terminal loams carries a soil horizon indicative of a prolonged slowing down or complete halt of deposition. Terraces formed at the same time show that during such periods of quiescence the stream channels cut into the underlying deposits. When the streams are incised as they are at the present time throughout most of the southern Argolid, and in the Mediterranean in general (Vita-Finzi 1969), only exceptional floods will deposit loam on the alluvial plains, and it is only the coastal zone that continues to receive some sediment.

Observations of semi-arid streams in action (e.g., Miall 1977, 1978; Patton and Schumm 1981; Picard and High 1973) and dating of ancient units show that alluviation in this climate zone is a rapid process. In the lower Ermioni valley, a modern sequence of stream-flood deposits and overbank loams has formed since about 1940; it has already reached a thickness that is about average for older alluvia in the area. The best chronologically controlled of any of the past alluvial units (the lower Flamboura; Figure 8) allows no more than a few hundred years for its formation. Thus, the several cycles of accelerated slope erosion and aggradation of streams in the southern Argolid appear to be but brief interruptions of a long history of slope stability, soil formation, and stream incision.

The long intervals of quiescence have their record only in the soils. Even where, as in the lower plains, upward growth of the overbank loam continued, the process was so slow and disturbance of the soil by plants, by animals, and later by plowing so intense that no

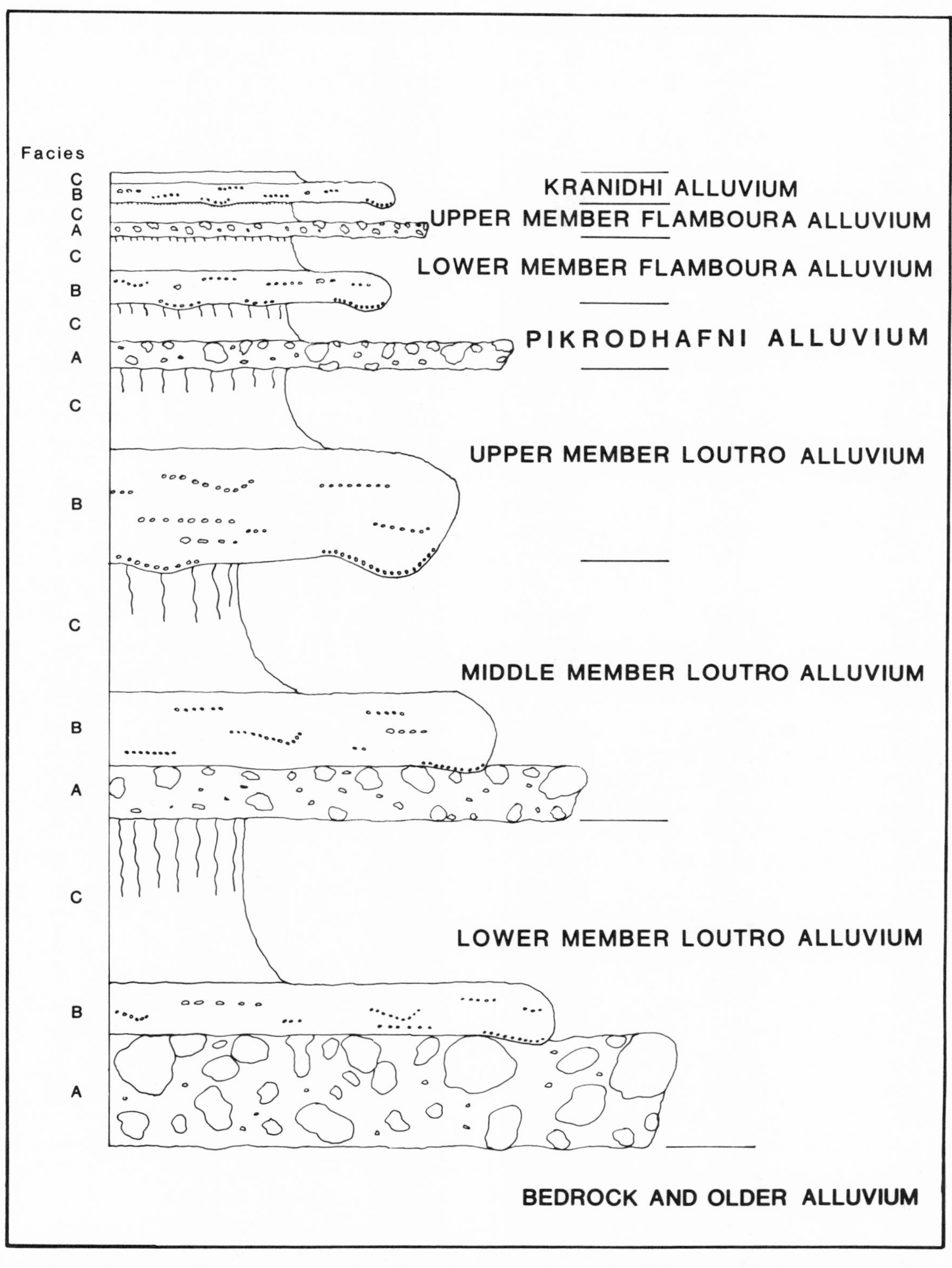

Figure 8. Late Quaternary stratigraphy of the southern Argolid. Capital letters on left correspond to facies type in Figure 7. The vertical axis, although not to scale, represents the relative average thicknesses of the units, and the horizontal scale is a measure of the degree of coarseness of the deposits. The length of the wavy vertical lines in the overbank loam facies of each unit is proportional to the degree of soil maturity (after Pope and van Andel 1984:Fig. 9).

record remains. Distinct soils, however, have formed in the loams of six of the seven stratigraphic units, as well as on suitable bedrock, mainly on the marls of the southern peninsula and on the ophiolites. In most cases, the A horizons of the soil profiles have been removed by erosion or by agriculture, but the B horizons of the five older ones and the pedogenic carbonates in their lower parts are quite resistant. The B horizon possesses properties that are useful for the determination of its relative age (Table 2), and sometimes sufficient to make an estimate of the time involved in producing the soil profile. Color intensification toward the red, increasing clay content, and an evolution from virtually structureless to columnar in several distinct steps all help to arrange the observed B horizons in a sequence of increasing age. The accumulation of calcium carbonate in the lower part of the B horizon also advances in several diagnostic steps. The rates of these processes, though dependent on precipitation, temperature, and substrate, are approximately known (Birkeland 1984; Harden 1982), and they support the statement made above that the intervals of stability and soil formation are very long compared to the brief times of slope wasting and valley aggradation. For example, the well developed soil that marks the terminal loam of the middle Loutro (Figure 8) indicates a minimum age of about 60,000 years for this alluvial unit (Pope et al. 1984).

THE LATE QUATERNARY RECORD

The oldest of the seven alluviation stages (Figure 8), the lower Loutro, dates to the Middle Pleistocene, some time between 250,000 and 300,000 years ago. A long break must have followed, because soil development on the middle Loutro (for which we have no isotopic date) suggests deposition around 60,000 years B.P. The upper Loutro is dated radiometrically on pedogenic carbonates at ca. 35,000 years B.P. (Pope et al. 1984). All are therefore Pleistocene in age, but their sedimentary facies are quite different. The lower Loutro is the only one in the area that has *grèzes litées* implying a climate colder than any that followed. This unit is dominated by loam with only very subordinate stream-flood deposits and has no debris flows, suggesting little runoff and soil erosion. The later Loutro alluvia were formed under less severe conditions. The middle phase has more stream-flood deposits, and the trend toward higher runoff and more slope erosion continues in the youngest member. Its stream-flood deposits fill deep scours in the underlying beds. Nowhere, however, is there evidence for extensive, catastrophic slope failures.

The next alluviation phase does not appear until approximately 4,000 years ago; there is no evidence for erosion or aggradation during the glacial maximum and the following deglaciation and postglacial climatic optimum. This Pikrodhafni alluvium is dated isotopically as at least 3,000 and more likely 4,500 years old (Pope et al. 1984). Associated artifacts of the Early Bronze Age and superimposed Middle Helladic sites and graves confirm the age estimate. This unit contrasts sharply with the preceding ones because it is dominated by debris-flow deposits. It is much less widespread than its predecessors, having been found only in the Fourni, Loutro, Ermioni-Iliokastro, and Pikrodhafni drainages, the principal areas of Early Bronze Age settlement (van Andel et al. 1986). Stream-flood deposits are rare, and the debris flows suggest large-scale sheet erosion and catastrophic deposition in the valleys, while the runoff remained essentially unchanged.

A period of quiescence followed that was long enough for the development of a distinct Pikrodhafni soil. It ended with the deposition of the lower Flamboura alluvium. This deposit is dated to ca. 300 B.C.-50 B.C. by associated late Classical and Classical/Hellenistic sherds and by superimposed Hellenistic and Early Roman sites. It is found in all drainages of the

TABLE 2

LATE QUATERNARY SOIL STRATIGRAPHY OF THE SOUTHERN ARGOLID
(van Andel and Pope 1984: Table 1)

Soil Name	*Horizons*	*Principal Diagnostic Features*	*$CaCO_3$ stage*
Kranidhi soil	A/C	A horizon thin to non-existant; parent-material stratification essentially unaltered.	none
Upper Flamboura soil	A/C	A horizon thin with faint crumb structure; some mixing of parent-material stratification in upper 25 cm.	none
Lower Flamboura soil	A/B/C	Well developed A horizon with dark chromas (high organic content); gradual horizon boundaries; cambic B horizon.	I
Pikrodhafni soil	A/B/C	Well developed A horizon; gradual horizon boundaries; distinct B horizon with 7.5-5 YR hues and clay films; carbonate nodules rare.	I-II
Upper Loutro soil	A/B/C	A horizon eroded; gradual horizon boundaries; 7.5 YR hues in B and C horizons; carbonate nodules common.	II
Middle Loutro soil	A/B/C	A horizon eroded; no stratification preserved in upper 2 m of fine-grained parent material; clear horizon boundaries; 5 YR hues in B and C horizons; thick Bt horizon.	II-III
Lower Loutro soil	A/B/C	A horizon eroded; no stratification preserved in upper 3 m of fine-grained parent material; clear horizon boundaries; 2.5 YR hues in prismatic Bt horizon.	III
Upland soil	A/B/C	Thick, well developed A horizon with coarse granular structure; thick Bt horizon.	?

southern Argolid and is characterized by stream-flood deposits. This indicates a concentration of runoff in gullies, due either to the widespread collapse of agricultural slope terraces or, less likely, to an increase in precipitation (van Andel et al. 1986).

Several centuries of stability produced an incipient soil in the lower Flamboura, but then another alluviation phase arrived which is rather weakly dated between ca. A.D. 600 and, at the latest, early modern time. Because its deposits are restricted to valleys below areas resettled in Middle Byzantine time and consist mainly of debris flows suggesting stripping of soils from cleared slopes, we prefer a date in the tenth to twelfth century A.D. (van Andel et al. 1986). Finally, the youngest of the alluvial units, the Kranidhi, is still in the process of formation, although its oldest parts in the Fourni valley may go back to the eighteenth century A.D. It is quite clearly the result of landscape destabilization due to recent changes in land use (Pope and van Andel 1984).

Complex as the local record is compared to Vita-Finzi's (1969) widely cited and applied scheme of Older and Younger Fills (e.g., Bintliff 1976a, 1976b, 1977), it may still be incomplete. Units may have been removed from the sequence by erosion or telescoped into a single phase. It is evident from Figure 8 that with decreasing age the units are thinner, in part as a result of a diminishing reservoir of slope mantle deposits as erosion outstripped weathering, but also because of better chronological resolution and better preservation. There seems to be little doubt that the sequence of the four younger units is reasonably complete and is separated from the Loutro aggradation by a very long time of no deposition, spanning many climatic changes. Furthermore, landscape destabilization seems to have been far more frequent in the last several millennia than during the whole Middle and Late Pleistocene.

The Argolid sequence thus shows that at least in this area the twofold scheme of Vita-Finzi and its narrow dating to late Pleistocene and post-Roman times are inadequate. However, although others have also argued that the recent alluviation history of the Aegean region was more complex than Vita Finzi implied (e.g., Davidson 1980; Raphael 1973; Wagstaff 1981), one should be on guard against applying the southern Argolid scheme elsewhere until more data are available.

CAUSES OF SOIL EROSION AND ALLUVIATION

The alternation of brief intervals of valley aggradation and long ones of slope stability and soil formation during the late Quaternary demands an explanation. In those cases where an increase in runoff and stream flow can be inferred from the deposits, a change in climate is a logical explanation. At other times, the sudden release of slope mantle in the form of debris flows presumes the existence of a thick weathered crust hitherto protected by a plant cover. Since even a few major deluges after some exceptionally dry years might dislodge such an accumulation, a true climatic change is not required in this situation.

Among the factors that are capable of inducing erosion and aggradation, changes in base level have always ranked high. Base-level changes include uplift and subsidence of the land as well as eustatic changes in sea level. They may be local, affecting only parts of a drainage basin, or they may alter geomorphic processes over a large region. Both tectonic and eustatic changes in base level have occurred during the late Quaternary in the southern Argolid, but there is no clear correspondence with the observed episodes of landscape stability and destabilization. The upper Loutro alluvium, for example, appears to have been formed contemporaneously with an intermediate level of the sea during a milder interstadial interval, but no corresponding alluviations accompanied equivalent changes in sea level earlier and later. Moreover, although tectonic subsidence has been claimed for the southern Argolid (e.g., Flemming 1968), there is no evidence that it was episodic nor that it resulted in changes in base level large enough to overwhelm those induced by global glaciations and deglaciations.

Climatic changes cannot be so easily dismissed. They might work either directly through changes in soil moisture or in runoff and stream flow, or indirectly through the plant cover. It has long been customary to correlate periods of soil formation or aggradation with major climate changes associated with glacial-interglacial or stadial-interstadial transitions (e.g., Dufaure et al. 1979), and more recently with the oceanic oxygen isotope record (Morrison 1976; Ponti et al. 1980), but the lack of chronological precision of both data sets continues to frustrate such attempts. Global climate changes have also been invoked to account for Holocene alluviations (e.g., Brakenridge 1980; Hassan 1981; Haynes 1968; Knox 1983; Wendland 1982), but the results are inconclusive, and local conditions and factors other than climate might be called upon with equal justification. Some of the proposed global alluviation phases broadly coincide with the Pikrodhafni and lower Flamboura alluvia, but the uncertainties involved are very large.

Moreover, these relatively simple correspondences do not stand up well in the light of recent advances in the study of fluvial systems which point to complex responses to single factors (Schumm 1977). In small watersheds in a subhumid or semiarid climate, for example, an increase in rainfall will benefit the vegetation and so counteract the erosion expected from the rise in runoff. The sediment yield diminishes rather than increases. In a more arid zone, the same increase in precipitation will not suffice to affect the plant cover, and both runoff and sediment yield rise. The impact of a climatic change or climate variation also depends on the season. In the southern Argolid, even a few years of heavy late summer rains would have a much larger impact on the sediment yield than an equivalent long-term increase in winter precipitation. A clear and unambiguous climatic signal is thus unlikely to emerge from alluvial sequences.

Nevertheless, the record provides some valuable information. The lower Loutro alluvium has a most likely radiometric date of 272,000 +52,000/−37,000 years B.P. (Pope et al. 1984), and the implied cold climate agrees well with oxygen-isotopic stage 8 (245,000-280,000 B.P.: Morley and Hays 1981). According to its facies, runoff was low and erosion only occasionally vigorous, pointing to a climate that was dry as well as cold.

The following cold and warm isotopic stages are either not represented or have not been recognized. The middle Loutro, as old as 60,000 years, was also the product of a cold phase, although not as cold as its predecessor, because we have found no trace of frost spall gravels. It is best placed in isotopic stage 4 between 73,000 and 60,000 B.P. It is of special interest because Middle Paleolithic (Mousterian) sites are associated with the soil profile at its top. Mousterian tools buried in this soil are encrusted with groundwater carbonate which was dated at 52,000 ± 13,000 B.P. (Pope et al. 1984). This places the Mousterian sites during the warmer isotopic stage 3 between ca. 40,000 and 55,000 B.P, or perhaps a little older (though not much, given the probable early onset of carbonate deposition).

Somewhat more firmly dated, within oxygen-isotope stage 3, is the upper Loutro alluvium which has a most probable radiometric age of 42,000 B.P. (minimum 35,000 B.P.: Pope et al. 1984). This was a mild period, and the sediments imply vigorous slope erosion with some debris flows.

It is curious that no alluvial deposits represent the subsequent high glacial nor the drastic climatic changes that accompanied the arrival of the deglaciation and, later, the warm and moist early and middle Holocene. There is no reason why any record of significant aggradation in this relatively recent interval should have been lost, and the gap therefore seems real.

The Pikrodhafni marks a turn to a drastically different geomorphic regime some 4,500 or 4,000 years ago. This change coincides with the first wide spreading of human settlement across the landscape of the southern Argolid, and we have argued elsewhere that human activity was the primary cause (van Andel et al. 1986). Late Paleolithic and Mesolithic hunter-gatherers may have burned the woods, and the land may have been modified to some extent by Neolithic farmers and pastoralists, but their sites are very few, and their activities have left no trace in the alluviation record. On the other hand, the many changes in the landscape since the Final Neolithic and Early Bronze Age have coincided so suggestively with specific episodes of human history that man's role as an agent, though perhaps not the only agent, can hardly be doubted. However, this part of the landscape history of the southern Argolid falls outside the scope of the Franchthi reports, and has been dealt with elsewhere (van Andel et al. 1986; Jameson et al., forthcoming).

It is clear that we have not been able to bring the search for causes of landscape destabilization during the late Quaternary to a fully satisfactory conclusion. This is likely to remain so until the nature and chronology of the climatic history of Greece and the chronology of its Pleistocene and early Holocene alluvial sequence have been given greater precision.

CHAPTER THREE
The Adjacent Sea

Franchthi Cave is at the present time situated in an eminently maritime position; just a few steps from the cave mouth, down a short rubble slope, is the seashore. This, however, is deceiving; for the larger part of its known occupational history, from ca. 25,000 to 5,000 years B.P., the shore was farther, much farther, away. In this chapter we explore the impact of late Quaternary changes in sea level on the Franchthi environment and its resources. The pertinent data have been obtained mainly during two brief marine geophysical cruises, a preliminary one in Kiladha Bay in 1979 (van Andel et al. 1980), and a more extensive survey in 1982, which covered the entire shelf of the southern Argolid including the Franchthi embayment (van Andel and Lianos 1983, 1984). For details regarding the seismic reflection method, equipment, navigation, and the actual data the reader is referred to the original publications.

LATE QUATERNARY SEA LEVEL HISTORY

Over the past 125,000 years, global sea level has fluctuated greatly (Figure 9), mainly because of changes in the volume of water stored in continental ice caps. Vertical movements of land and seafloor, if they contributed at all, did so only to a small degree, because tectonic rates are limited to a few centimeters/1,000 years by the slow rates of thermal contraction and expansion or mantle flow. Faulting in grabens such as the Gulf of Corinth may achieve higher rates over the intermediate term, but this does not apply in the southern Argolid. The elevations of Neotyrrhenian shores in the upper Gulf of Argos (Kelletat et al. 1976), around southeastern Lakonia (Kelletat et al. 1976; Kowalczyk 1977), and near Elaea in the northeastern Gulf of Lakonia (personal observation) are from 1 m to 5 m above present sea level, in good accord with widely accepted values for sea levels of the last interglacial (Cronin 1983). In the southern Argolid, the lowest sea level at ca. −120 (van Andel and Lianos 1983), near the best global values, leaves little room for major long-term tectonic subsidence such as the 2-3 m/1,000 yrs suggested for this area by Flemming (1968, 1978).

During the early part of the last glacial (oxygen isotope stage 5), sea level fluctuated considerably between highs of ca. 10-20 m below present sea level and unknown low stands (Figure 9). Assuming that the oxygen isotope curve reflects to a first approximation the removal of water from the sea and its storage in ice caps, the lows cannot have been as deep as during the glacial maximum. Major cooling began around 80,000 B.P., a time we may regard as the true beginning of the last glacial (Kukla and Briskin 1983), and the sea sank to a low, but again not well defined, level. During isotope stage 3 the climate improved somewhat, with sea level between 30 m and 40 m below present, until a major expansion of ice caps beginning around 30,000 B.P. caused the sea to fall to its lowest position of the entire glacial, well defined in the southern Argolid by the upper edge of typical marine slope deposits (van Andel and Lianos 1983).

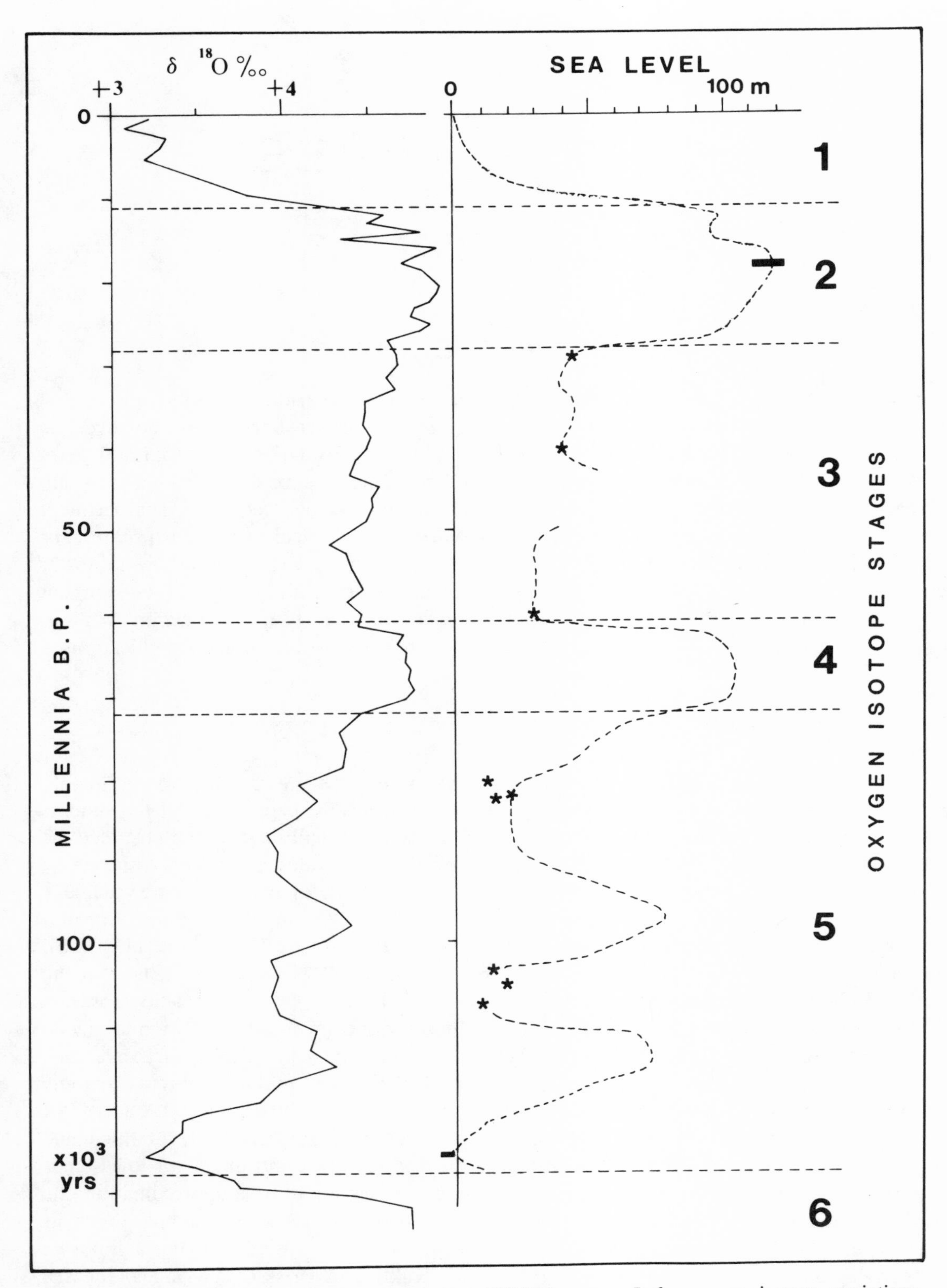

Figure 9. Global sea-level changes of the past 125,000 years. *Left:* oxygen isotope variations with time based on benthic deep-sea Foraminifera (N. J. Shackleton 1977). *Right:* sea-level changes. Stars mark high stands estimated from raised beaches on Barbados, New Guinea, and Haiti (Bloom et al. 1974; Chappell 1974; Dodge et al. 1983; Matthews 1973). Bars show deepest late glacial level in the southern Argolid and highest global stand during last interglacial. Dashed line is the sea-level curve. Large numbers indicate oxygen isotopic stages, dated after Morley and Hays (1981).

When the climate began to improve and the ice to melt, the sea rose. Data on the influx of meltwater in the Gulf of Mexico (Kennett and N. J. Shackleton 1975) and other paleoceanographic indicators (Berger 1977; Berger and Killingley 1982; Delmas et al. 1980; Johnson 1980) have placed the initial rise between ca. 16,000 and 14,000 years ago, but the oxygen isotope data of Pastouret et al. (1978) put it somewhat later.

The details of subsequent events are still controversial. The first warming probably led to a significant thinning of the ice caps and a marked northward retreat of glacial conditions in the North Atlantic (Ruddiman and McIntyre 1981a, 1981b), and converted about one third of the ice volume into seawater (Duplessy et al. 1981). Colder conditions then returned for a while leading to a readvance of the ice, a southward march of the tundra in northwestern Europe, and a return to a glacial North Atlantic. The final stage of melting began about 10,000 years ago, the postglacial climatic optimum was reached ca. 7,000 B.P., and the rise of the sea slowed down markedly.

At the present time, there is reasonable agreement that the deglaciation and associated rapid sea-level rise did indeed take place in two steps (Ruddiman and Duplessy 1985), but the dating of the first step remains controversial. Duplessy et al. (1981) placed it between 16,000 and 13,000 B.P., followed by a long pause represented by the cold stages of the European Older and Younger Dryas. Mix and Ruddiman (1984) date the first step between 13,000 and 11,000, on the basis of core data from the equatorial Atlantic, whereas Berger et al. (1985) put the entire first melt and rise just after 14,000 B.P. and before 13,000 B.P. Currently available sea level data (Bloom 1977, 1983) accommodate the two-step rise, but are open to various interpretations regarding their timing.

Whatever may have been the precise start, a rise of the sea as rapid as 5 cm/year was underway around 13,000 B.P. At this rate, in a very short time much coastal land was lost on wide shelves such as the Adriatic, an environmental catastrophe that could not have escaped the attention of those present to witness it.

The rise of the sea continued, albeit much more slowly, even after the northern ice caps had fully melted away, and the equivalent of approximately 20 m of water has been added to the oceans over the last 7,000 years, without counting the isostatic compensation for the extra load. The source was probably the West Antarctic ice sheet, generally thought to be unstable, and the rise continues at the present time, amounting to a few tens of centimeters per century according to the best tide-gauge data (IAPSO 1985).

The course of the rise of the postglacial sea was not merely determined by the release of water from melting icecaps, but also by the isostatic response of once ice-covered continents to the removal of the load and that of the ocean floors and continental margins to the weight of the added water (Clark et al. 1978; Peltier 1980). The gravitational attractions of large ice and water introduced additional local and regional complications. It is now quite clear that, except in the broadest sense of the word, no globally valid eustatic sea level curve shall be discovered (see Kidson 1982 for a review of the subject).

Even regarding the most general course of events, a number of views exists. There are those who interpret the record in the simplest possible way as a steady rise at an ever-decreasing rate, present level having been reached some time during the past few thousand years. Others, for example Curray (1960) or Fairbridge (1961; for a summary, see Kidson 1982), regard the evidence as forcing the assumption of a series of stillstands or reversals which interrupted the postglacial rise, and perhaps included levels slightly above the present one during later phases. Such stillstands or brief reversals might have been induced by cold "little ice ages" peaking around 8,000, 5,300, 2,800, and 200-300 B.P. which have been recognized from advances of European and North American mountain glaciers (Denton and Karlén 1973).

A local sea level curve is therefore needed for the southern Argolid. Unfortunately, data for the construction of such a curve are scarce, deriving mainly from the submerged temples, stadium, city walls, and harbor of Halieis (Jameson et al., forthcoming: Chapter 3.3). Recently, two dates have been obtained from charcoal in cores taken at a submerged Neolithic site a few hundred meters offshore from Franchthi Cave (Gifford 1983). These ages of 6,217 ± 129 and 7,612 ± 148 radiocarbon years (J. A. Gifford, personal communication, November, 1985) indicate that between ca. 6,500 and 5,000 B.C. sea level off Franchthi Cave was at least 11 m lower than it is today. [For additional dates from offshore coring, see Jacobsen and Farrand 1987.—EDITOR]

No earlier local dates are available and, though recognizing the uncertainties and error margins of the procedure, we are forced to estimate early postglacial sea levels from a "global eustatic" curve as used by van Andel and Lianos (1983, 1984) and J. C. Shackleton and van Andel (1986). It was obtained by drawing a curve through the center of density of the cluster of radiocarbon dates available for those continental margins of the world not seriously affected by either isostatic compensation or tectonics (van Andel and Lianos 1983, 1984). Modified to accommodate the now widely accepted two-step rise, the local and general curves appear in Figure 10. The lowest stand shown is the one documented for the area itself; the sea remained at approximately this position from shortly after 30,000 B.P. to as late as about 14,000 years ago.

PLEISTOCENE AND HOLOCENE SHORES AT FRANCHTHI

The streams draining the southern Argolid are generally small and so are their sediment loads. As a result, the shores of this region have prograded little during the Holocene, and the history of the Argolid coast during the postglacial sea-level rise lies concealed mainly on its continental shelves. The approach used by Kraft and co-workers to outline the history of various Greek coastal zones (Kraft et al. 1975, 1977, 1980) has therefore little application here. Instead, we have turned to marine geophysical methods to obtain the data needed for a reconstruction of the position and aspect of the coasts of the Franchthi embayment in the past.

The principal technique used was subbottom seismic reflection profiling, a modified form of echosounding in which a strong acoustic signal of low frequency penetrates the bottom and is reflected by subsurface discontinuities. The record resembles a geological cross section, showing stratification, attitudes of beds, and sometimes the nature of the deposits (e.g., Curray and Moore 1963; van Andel and Sachs 1964), with echo time and ship's travel time as vertical and horizontal scales. Knowledge of the sound speed and the ship's positions permits their conversion to true depth and distance. A side-looking sonar provides information regarding the nature of the bottom over ca. 150 m to either side of the ship's track. Additional technical information can be found in van Andel and Lianos (1983).

The deepest reflector seen on the Argolid marine records is usually also the most prominent one. At times it is broad and ragged-looking (Plate 4: A), indicating a rough, rather poorly reflecting surface. Near the shore, this surface merges with outcropping bedrock and therefore probably represents its buried continuation. More common is a thin, sharp, dark basal reflector (Plate 4: B). Its smoothness and level aspect across most of the Franchthi embayment, and its dissection by stream channels suggest a former subaerial plain. In Kiladha Bay, this reflector has been traced to its equivalent on land (van Andel et al. 1980), a red, gravelly loam of late Pleistocene age (upper Loutro alluvium), and coring offshore in the bay has yielded a shallow marine sand immediately above it (Gifford 1983). This basal

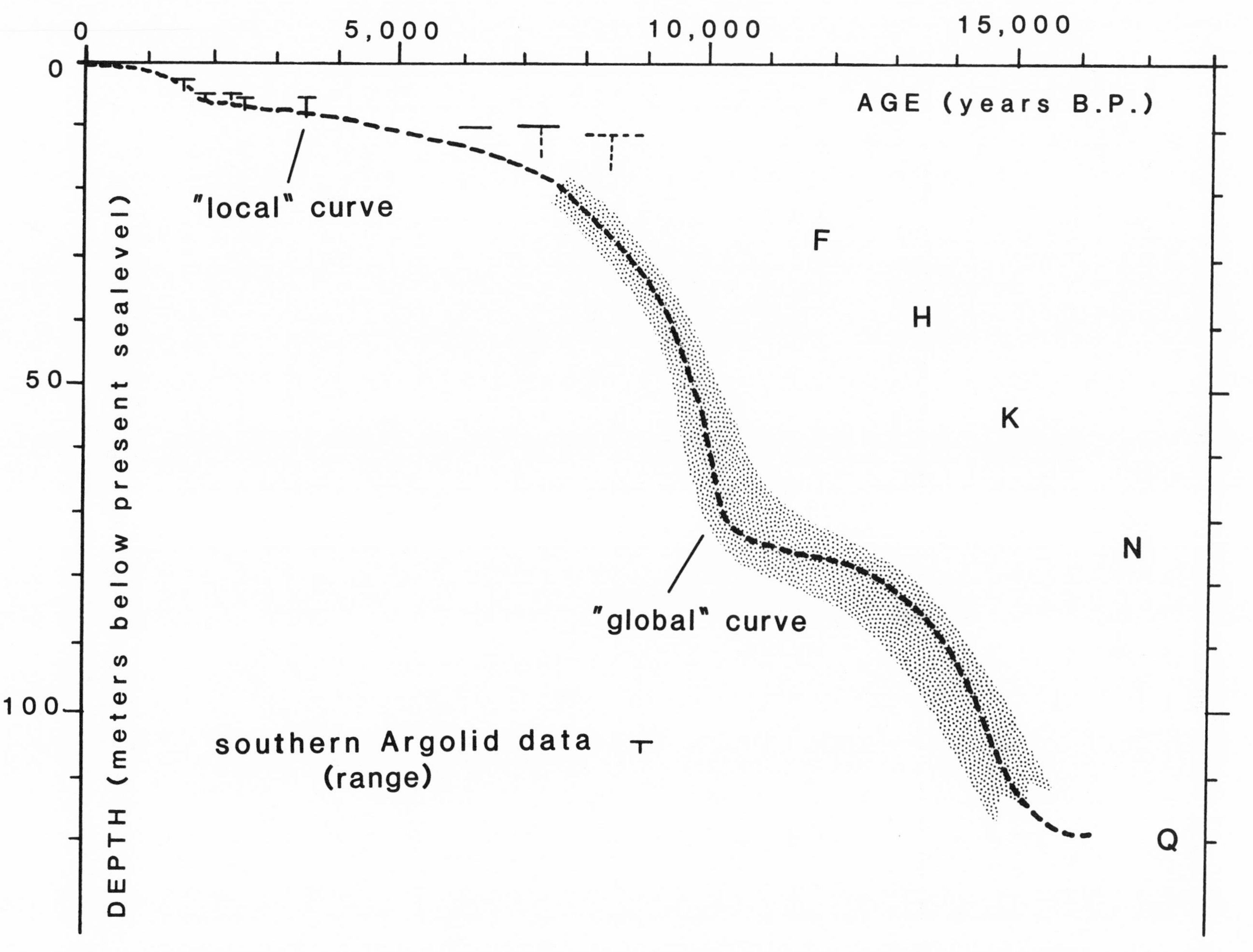

Figure 10. Postglacial sea-level rise in the southern Argolid. Horizontal bars in ''local'' curve indicate time range of dates, vertical bars show that sea level was below elevation at which date was obtained. Labeled arrows indicate shore clusters of Figure 11, also used for the construction of Figures 13-17. For an explanation of the ''global'' curve, see text.

reflector may therefore be regarded as the surface of the late Pleistocene and early Holocene coastal plain.

Here and there this buried surface is marked by minor irregularities. Among them, small scarps ranging from 1 to 3 m in height are most common (Plate 5: D, E). They always face seaward and usually have a flattened area or slight swale at their base. Also common are low ridges, generally less than 1 m in height and with widths varying from 10 m to more than 50 m (Plate 5: F, G). The smaller ones resemble in shape and size the present beaches of the peninsula, for example the small nearshore bars in Potokia and Kranidhi bays on the east side. The larger ones are similar to the gravel barriers that separate the Thermisi and Ververonda lagoons from the sea.

Many channels are cut in the basal surface, some narrow and V-shaped, others up to a few hundred meters wide (Plate 5: G; Plate 6: G-L). The channel walls tend to be too steep (<12-15°) to reflect sound, and their floors are always obscured by distinctive, finely stratified, very fine-grained sediments that may have been deposited when the rising sea flooded the channel mouths. In the Franchthi embayment, these channels connect to form the lower, now submerged portions of a sizable drainage system.

At a depth of 118-120 m below sea level, the basal reflector terminates against a wedge of fan-shaped strata identical in aspect with the silty clays of most continental slopes (Plate 4: C). No basal reflector can be seen underneath although sound penetration is excellent. One assumes with confidence that the boundary between the basal reflector and the slope deposits represents the lowest sea level of the late Quaternary.

Above the basal surface lies a blanket of transgressive deposits commonly less than 2 m thick. Only in sheltered places such as Kiladha Bay where sediments are trapped does this layer reach up to 5 m in thickness (van Andel and Lianos 1983: Table 1). The acoustically transparent post-transgressive deposits show little stratification except in channels and appear to consist of very fine sediments laid down in quiet water well beyond the shore. Given the high rate of shoreline migration associated with the rapid early rise of the sea, it is quite probable that any given spot moved beyond the nearshore zone before a measurable deposit of coarser, stratified coastal sediment could form. Only in depths shallower than about 30 m does stratification appear in Holocene marine deposits. Samples from this zone consist of fine, silty and shelly sands and sandy silts indicative of the more enduring nearshore conditions of the later, slower part of the transgression.

The scarps and ridges, so similar to the coastal features of the area today, are interpreted as the remains of shores formed during the glacial lowstand of the sea and the subsequent transgression. From the bases of the scarps and the tops of the ridges the positions of the corresponding sea levels can be determined with a precision of 1-2 m. These positions cluster around a number of discrete depths rather than being randomly distributed (Figure 11). The clusters imply that the rise of the sea was not continuous but interrupted by brief still-stands or reversals which allowed the shore, at other times moving too rapidly, to leave its mark on the plain. Below −40 m the clusters are widely spaced and distinct, because the rise of the sea between stillstands was rapid. At shallower depths the spacing is closer and the assignment of some features to one or the other of adjacent clusters may be arbitrary. Above −10 m seismic profiler data are too sparse.

There is no independent information, for instance from cores, to confirm that the features tabulated in Figure 11 are indeed old shores, but the cumulative evidence is persuasive. Neither do we have direct dates for them and, though reluctantly because of the broad error margins, we must make use of the sea level curve of Figure 10 to estimate the ages of the various main shores. Fortunately, even for the older ones, the steepness of the curve prevents the error margins from becoming too uncomfortably broad.

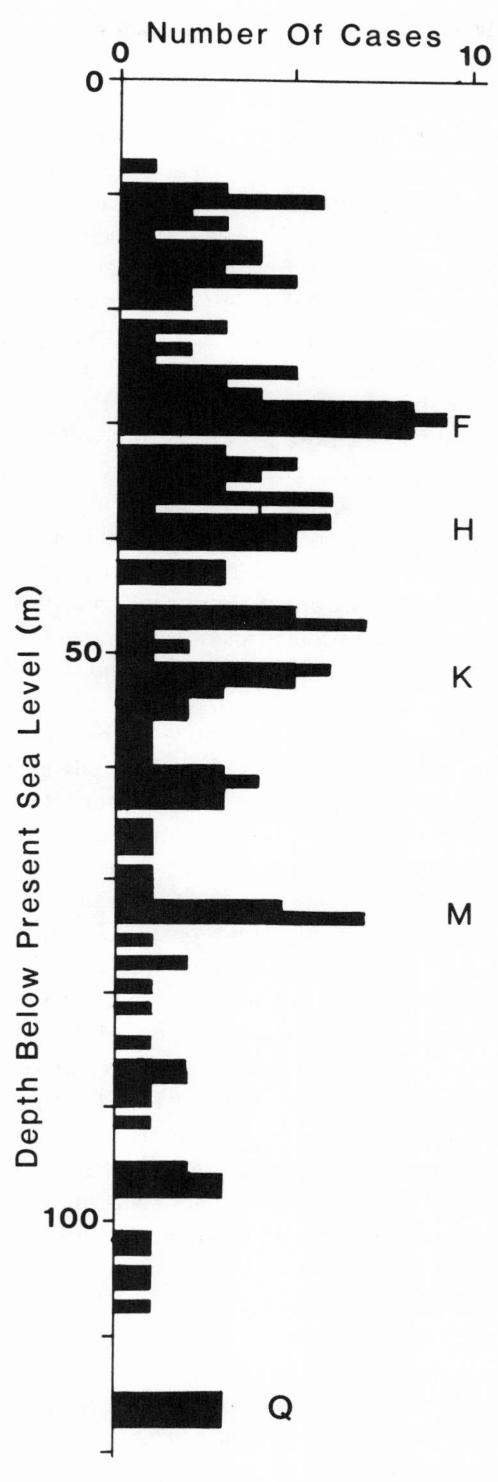

Figure 11. Clustering with depth of postglacial shore features in the southern Argolid. After van Andel and Lianos (1983:Fig. 9).

The geophysical data discussed briefly above, together with analogies drawn from modern shores of the Peloponnese, with guidance from general principles of coastal geomorphology (e.g., Curray 1964; Davis 1978: various articles; Kraft 1985; Pethick 1984; Wright and Thom 1977), permit us to reconstruct in some detail the coastal environments of the past in the Franchthi embayment (J. C. Shackleton and van Andel 1986).

The present coast of the Franchthi embayment is in the main rocky (Figure 12). At the foot of the limestone range in the north, scarps a few meters high fall steeply and usually without a beach to a rocky seafloor a few meters deep. In the south the bedrock is softer and coastal retreat has produced taller, actively eroding cliffs, usually but not always fronted with narrow cobble or shingle beaches. Beaches of fine gravel or sand are rare, but they do occur locally north of the Franchthi and south of the Kiladha headland. The only mudflats of the area lie protected in the inner part of Kiladha Bay. Offshore, the bottom consists of shelly, silty fine sands grading outward to a silty mud. The gravel beaches generally have been built out over a sand base. Only near some of the rocky capes does the bottom consist of bedrock to a depth of 5-15 m.

Only a little sediment is delivered by streams to the sea, mainly to Kiladha Bay where it is trapped. Thus, the variety of coastal depositional environments is limited. Elsewhere in the southern Argolid where the sediment supply is more prolific, depositional environments exist that complement those found in the Franchthi area, such as the barrier-lagoon complexes on the southwest and east sides of the peninsula and the small torrential deltas east of Thermisi (Figure 1).

During the last glacial maximum, the entire embayment was land, terminating at its western shore in beaches, marshes, and a few low cliffs strung nearly straight between two rocky promontories (Figure 13). The extensive coastal plain, about 6-7 km wide, was traversed by the Franchthi river of which today only disconnected upper tributaries remain, and by several southern streams. At the southwestern corner lay a large, muddy bay, sheltered behind a rocky point.

The first of the two stages of the rapid postglacial transgression has left few traces, but a broad coastal zone of marshes, channels, inlets, and beaches formed when the sea had reached a level of about −73 m (Figure 14; stage M of Figure 11). Its well-developed features suggest a prolonged halt or very slow rise of the sea, and the assignment of this stage to the pause between the two deglaciation steps, to the cold Dryas stadials, appears justified. This places the complex no later than 11,000 B.P. and probably no earlier than 13,000 B.P. The coast was still straight between now somewhat extended rocky shores in the north and south, but the sheltered southwestern bay with its promontory had disappeared. It was now 4 km from the shore to the cave. Two small, highly reflective mounds near an inlet may possibly be shell middens (Plate 6: L; see also J. C. Shackleton and van Andel 1986).

The subsequent rise of the sea was once again rapid and left few traces except for some small scarps cut into the quickly shrinking coastal plain, thus bearing witness to a fast-moving transgression and a lack of sediment supply (Figure 15).

When the transgression, about to slow down, reached the gentler upper part of the shelf, a distinct change in coastal development occurred (Figure 16). Between the still lengthening cliff coasts in the north and south one finds once more a depositional zone of beaches, bars, and marshes or salt flats on both sides of a large inlet. This inlet, pointed toward Franchthi, now a mere 2 km from the present shore, constituted the beginning of what is today Kiladha Bay. In the inlet and for a fair distance offshore, the water was very shallow, less than 2 m deep, and the bottom consisted of mud or fine silt, a situation much like the upper Gulf of Argos today.

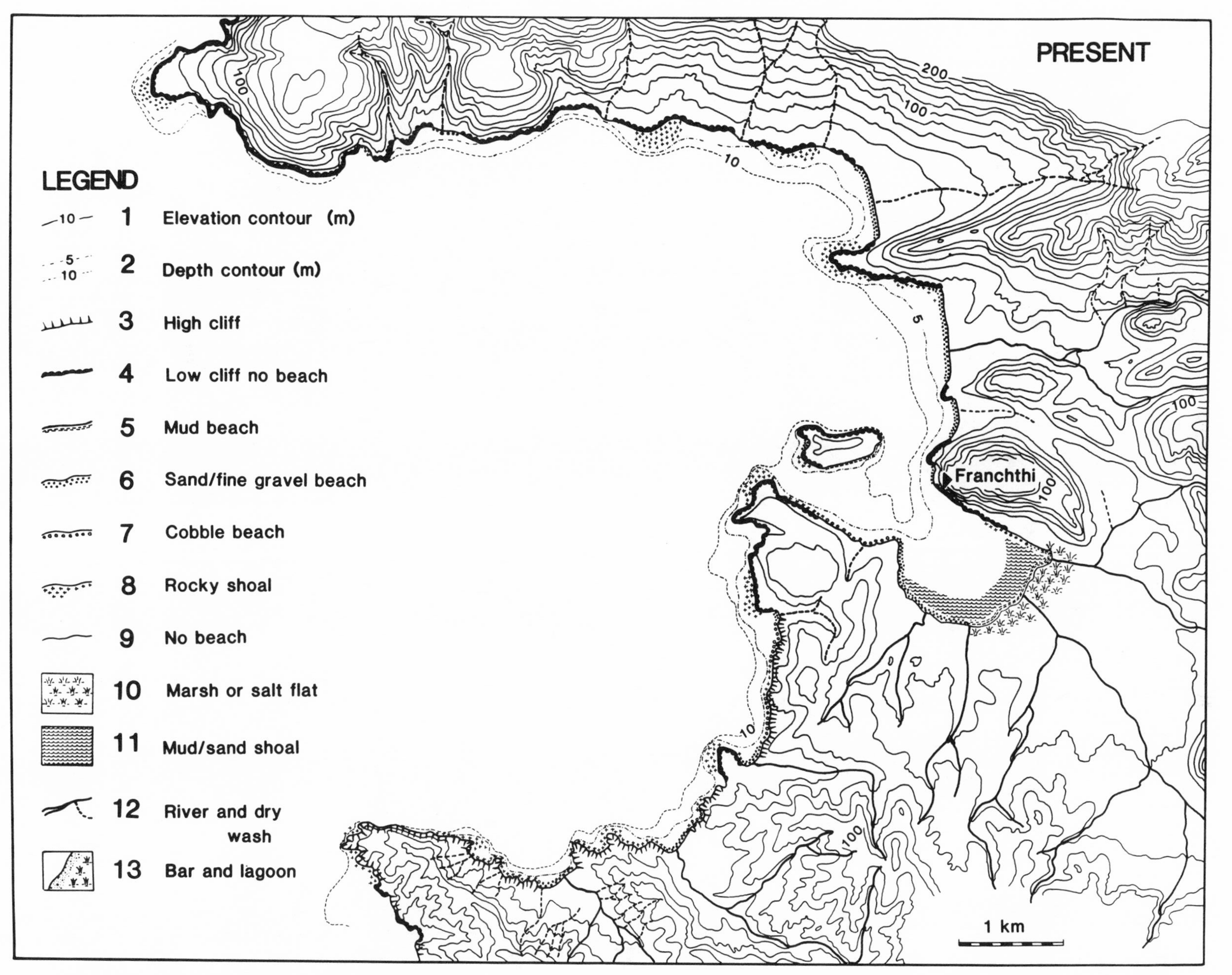

Figure 12. Present shores of the Franchthi embayment. Elevation contours (in meters above present sea level) and the 5 and 10 m isobaths are from Greek topographic maps, scale 1:50,000 and 1:5,000. The legend also applies to Figures 13-17.

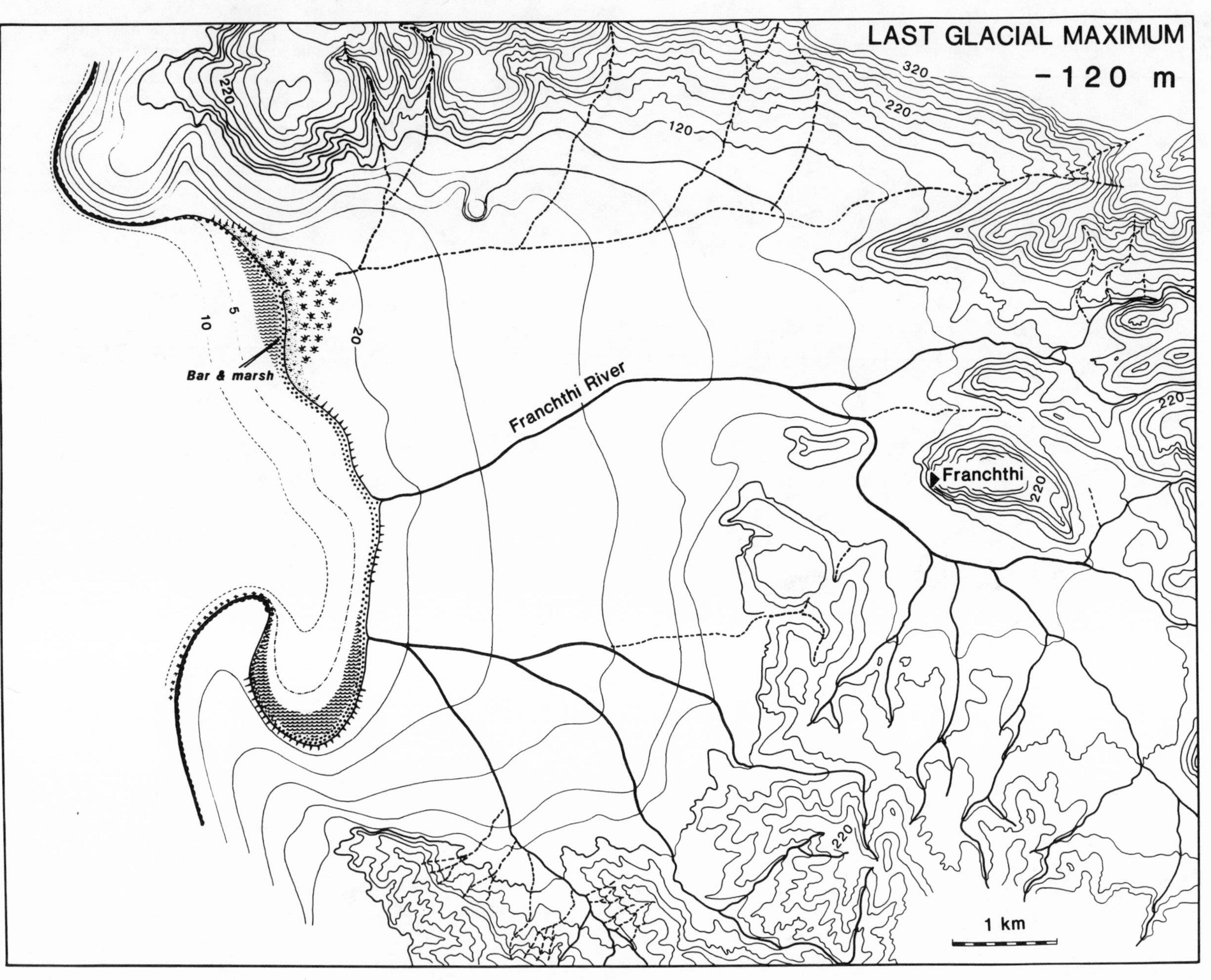

Figure 13. Shores of the Franchthi embayment during the last glacial maximum. Elevation contours labeled in meters above sea level at that time. Legend on Figure 12.

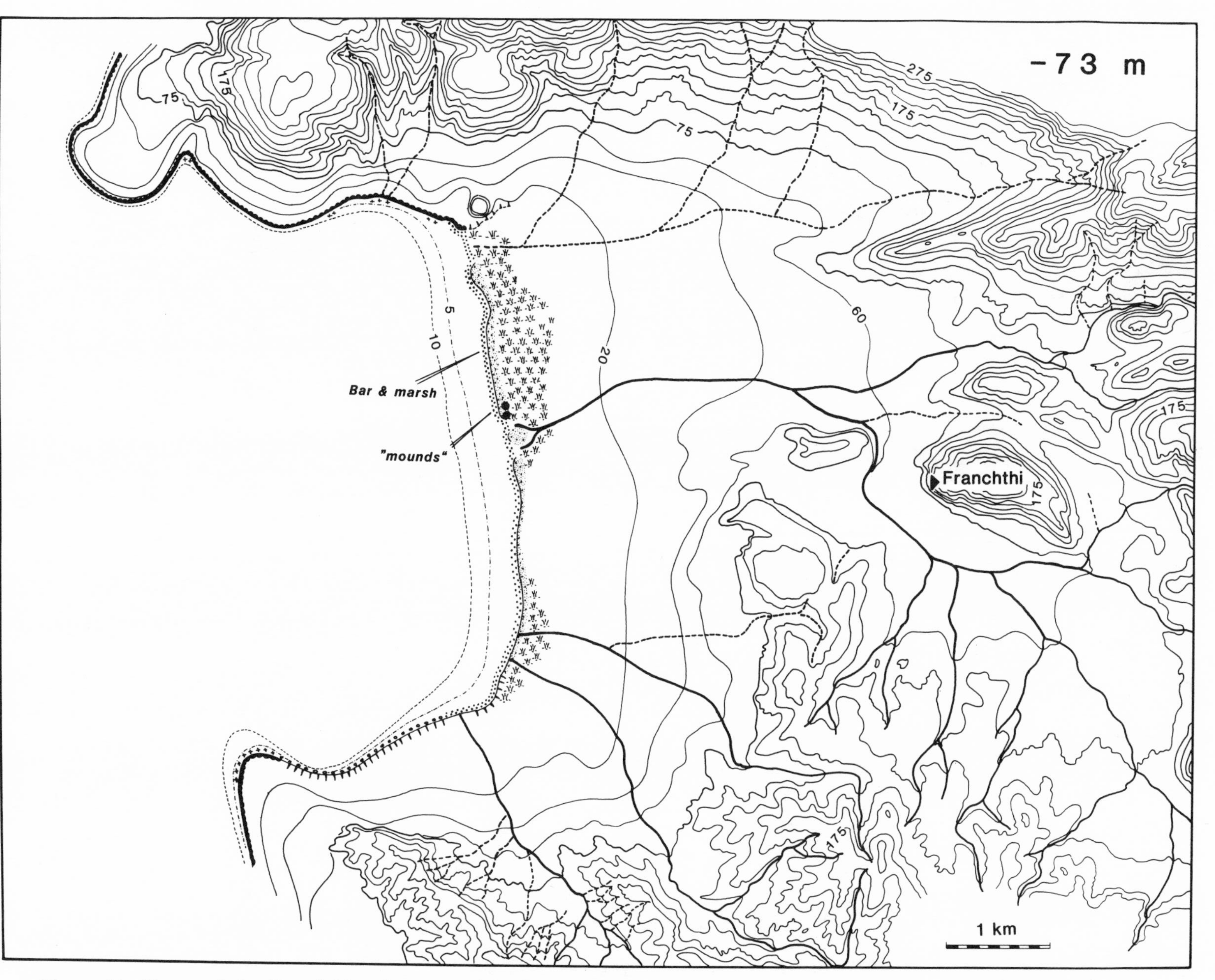

Figure 14. Shores of the Franchthi embayment when the sea stood at −73 m, probably between ca. 13,000 and 12,000 B.P. Elevation contours in meters above sea level at that time. Legend on Figure 12.

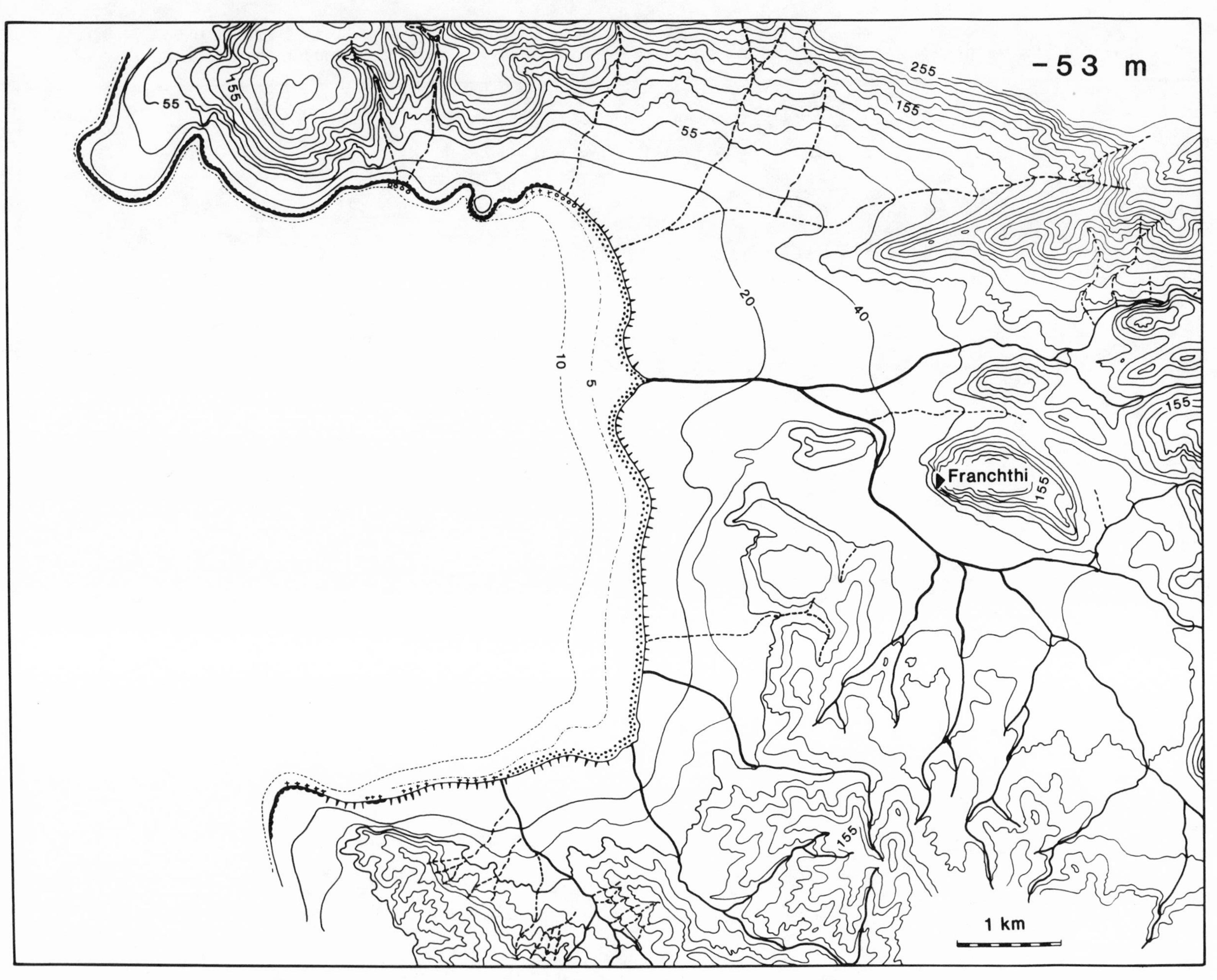

Figure 15. Shores of the Franchthi embayment when the sea rose past −53 m, probably around 11,000 or 10,000 B.P. Elevation contours in meters above sea level at that time. Legend on Figure 12.

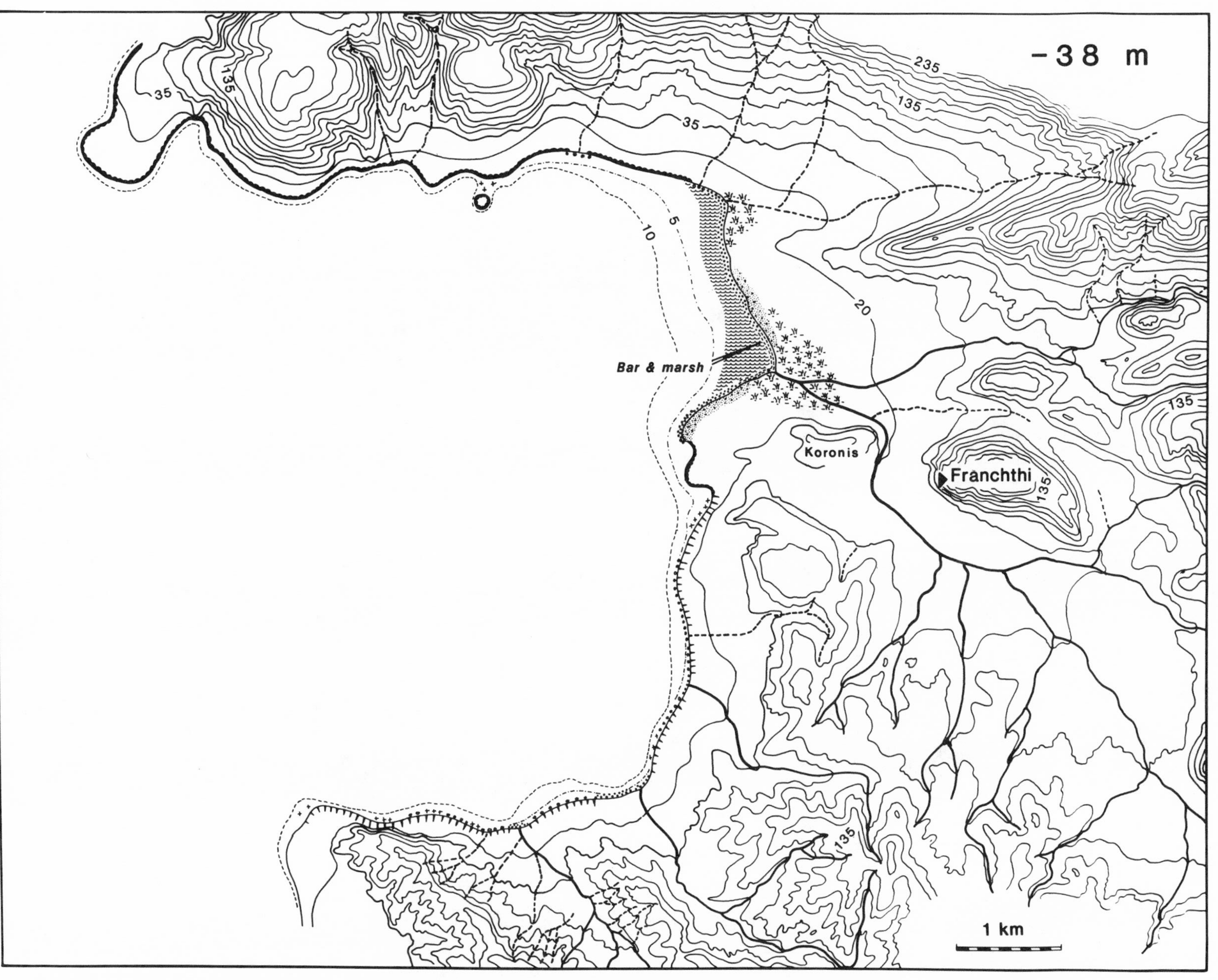

Figure 16. Shores of the Franchthi embayment when the sea stood at −38 m, probably about 9,500 or 9,000 B.P. Elevation contours in meters above sea level at that time. Legend on Figure 12.

As the sea rose and the inlet grew landward past the entrance to the cave, the marsh and beach complex shrank (Figure 17), and the coasts of the Franchthi embayment became much as they are today. Due to the trapping of fine sediment within the bay, sand beaches, marshes, and mudflats along the open sea disappeared, and rocky shores with gravel and cobble beaches dominated to either side of the bay entrance.

The evolution of the bay itself (van Andel et al. 1980) can be traced from a close grid of seismic reflection traverses obtained in 1979 (Figure 18). As on the shelf outside the bay, the late Pleistocene and early to middle Holocene subaerial surface forms a strong reflector beneath the later transparent deposits. This pre-transgressive surface has been sculpted into a set of broad, shallow stream channels converging on a curved, fairly deep inlet between Franchthi and Koronis Island.

The transgressive sediments resting on this old land surface contain numerous internal reflectors which can be cross-correlated (Figures 19 and 20: reflectors a-h). Each reflector marks a former sediment-water interface, an old bay floor, and so represents a time plane. The reflectors are concave upward where they delineate the channels and channel banks of the Franchthi inlet. Their edges meet the underlying basal surface at a low angle and at a depth roughly corresponding to sea level at the time of their formation. This implies that the depths of the contacts should remain constant throughout the bay, and indeed they do so, extending farther inland with decreasing age.

Assigning ages to the old bay floors is difficult. For the younger ones (d and later), there is a reasonable correspondence with alluviation phases recognized on land (Jameson et al., forthcoming: Chapter 3.3), but we can merely distribute the earlier ones (f, g, and h) rather arbitrarily over the Neolithic and earliest Bronze Age prior to the deposition of the Pikrodhafni alluvium around 2,500 B.C.

With sea level at −17 m (Figure 21), a narrow inlet, probably fringed with marsh, penetrated a short distance between Franchthi and Koronis Island. The time may have been around 7,500 B.P. As the sea continued to rise ever more slowly, the inlet advanced up the main stream channel, eventually dividing into several branches connected with the streams of the inner bay. Along the east bank the shore was flat, then rose gently toward the edge of a ca. 100 m-wide terrace of Loutro loam, watered by springs. It is on the edge of this terrace that an open-air Neolithic site, overlooking the inlet and a strip of marsh a few meters below, was found under 5.5 m of water and 4.5 m of mud (Gifford 1983; personal communication, 1985). The greater part of this terrace and the site were flooded when the sea rose to approximately 10 m below present level; our best guess is that this may have happened about 4,500 to 5,000 years ago.

At that time the bay had acquired virtually its present configuration except that Koronis Island remained attached to the mainland on the Kiladha side. Extensive marshes, of which little remains today, occupied the inner parts of the bay. Only minor changes have occurred since then, although the once-deep channels have now shoaled and the inner shore has somewhat advanced seaward (Jameson et al. forthcoming: Chapter 3.3).

THE SEA AROUND THE ARGOLID

Intensive study of ocean sediment cores during the last 20 years has greatly enhanced our understanding of the characteristics and circulation patterns of the oceans of the past. This has been largely due to the widespread use of stable isotope data of oxygen and carbon and of methods for extracting quantitative environmental data from microfossil assemblages (Cline and Hays 1976; Imbrie and Kipp 1971). These methods have also been applied to

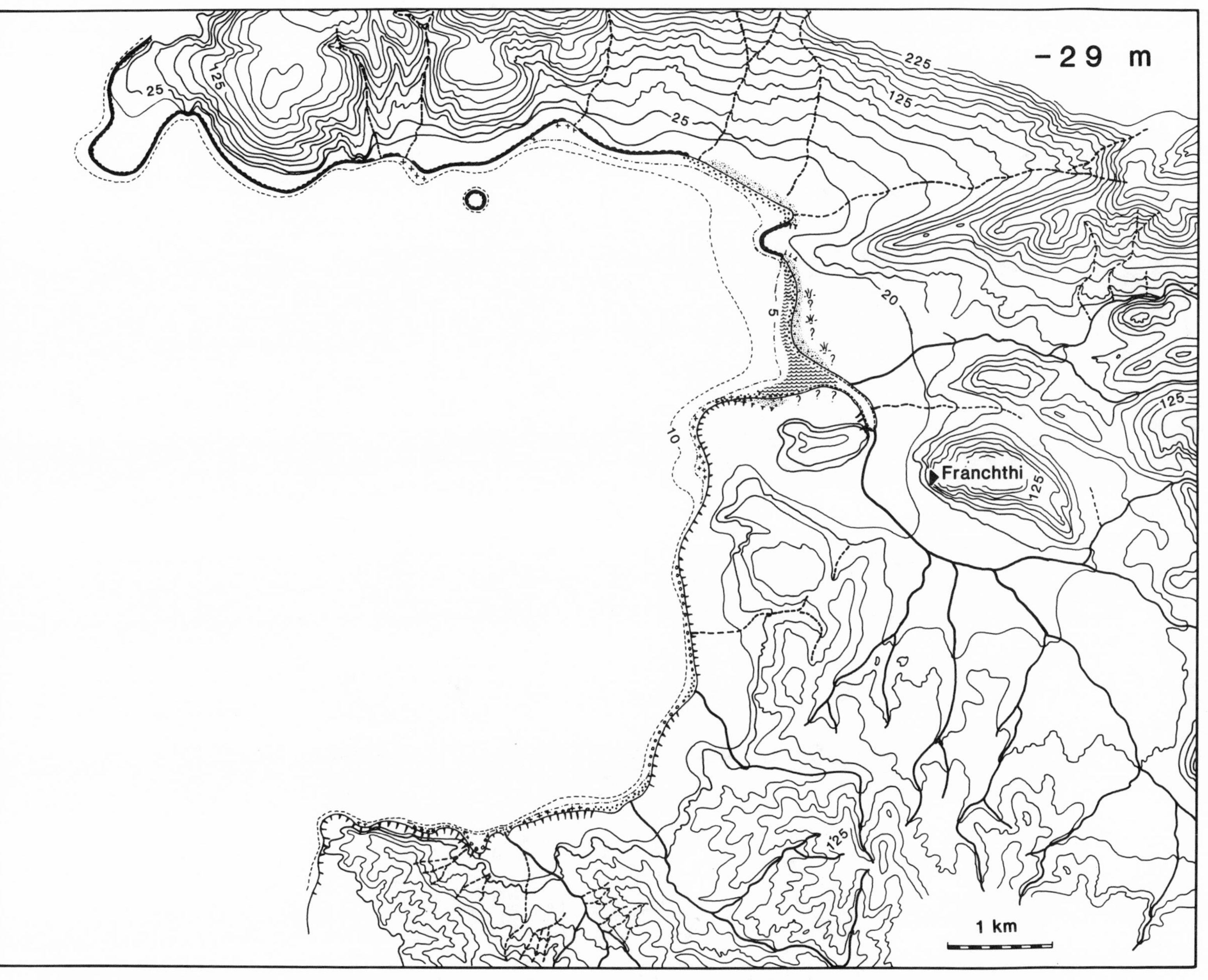

Figure 17. Shores of the Franchthi embayment when the sea stood at ca. −29 m, probably about 8,500 B.P. Elevation contours in meters above sea level at that time. Legend on Figure 12.

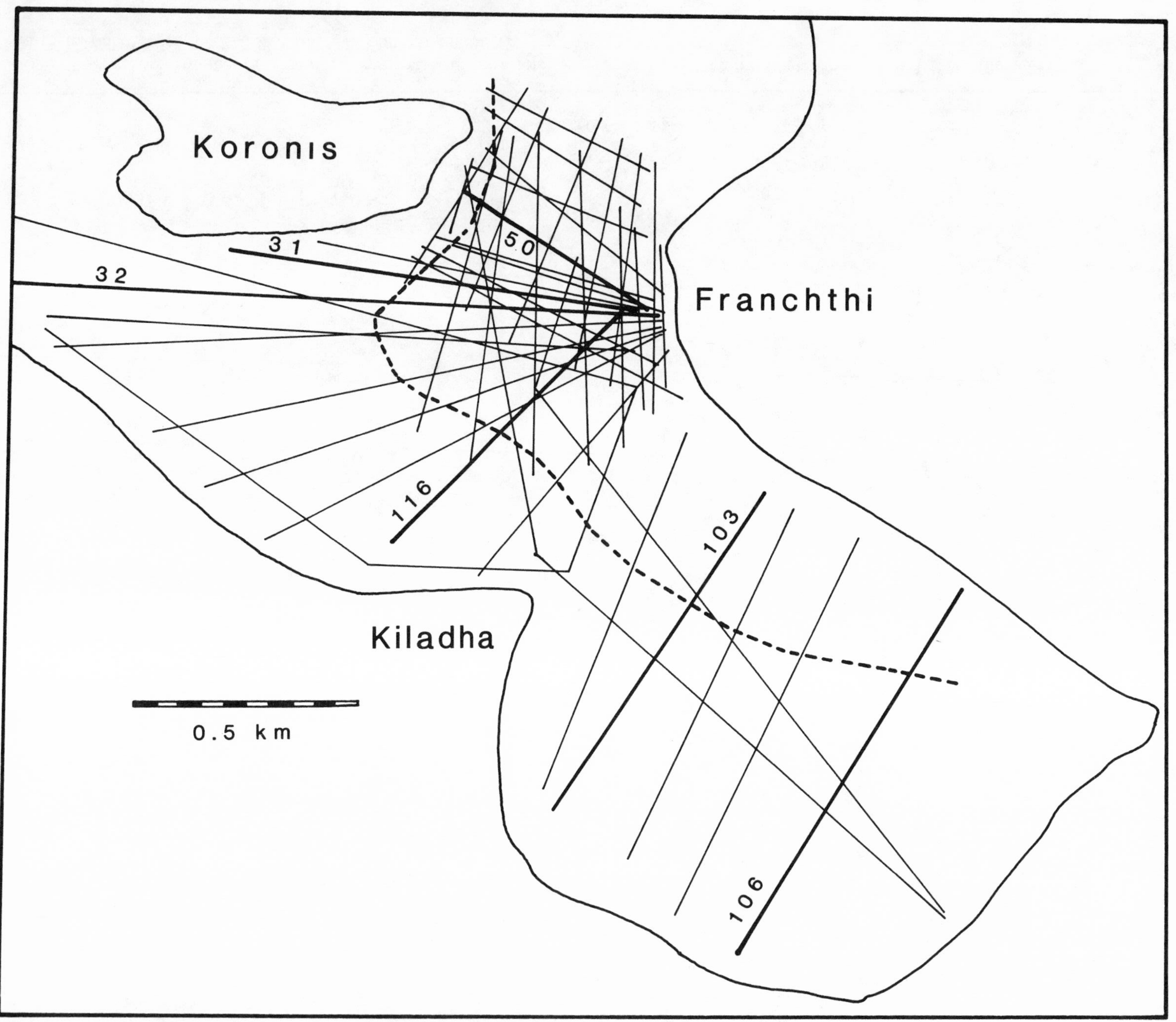

Figure 18. Geophysical survey tracks in Kiladha Bay (1979 survey: van Andel et al. 1980). Numbered solid lines are profiles shown in Figure 19. Heavy broken line is profile of Figure 20.

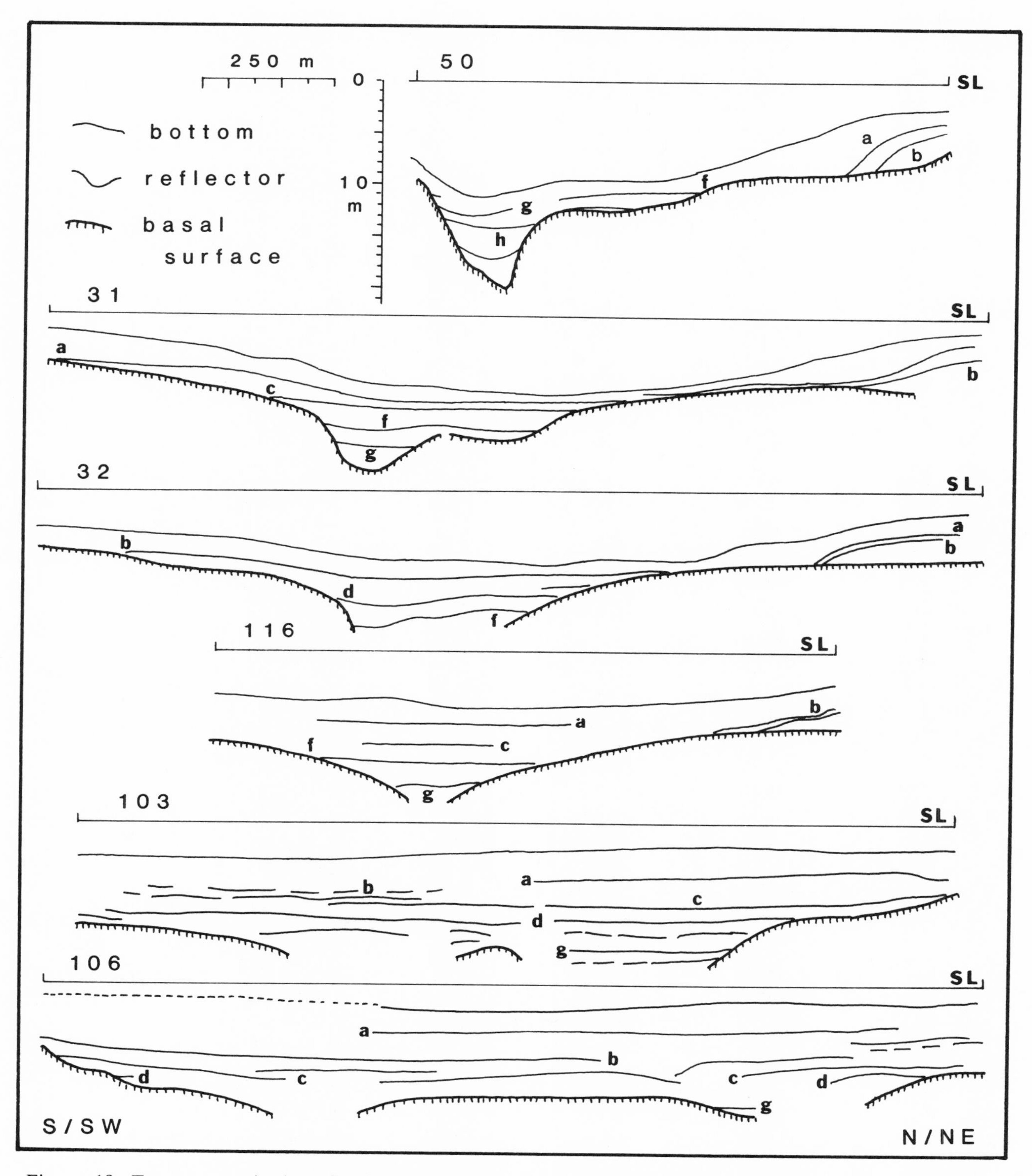

Figure 19. Transverse seismic reflection profiles of Kiladha Bay. Numbers correspond with locations in Figure 18. Profiles are arranged from west (*top*) to east (*bottom*). Thin labeled lines between hachured basal reflector and bottom represent bay floors during successive stages of the flooding and filling of the bay. Depths are below present sea level.

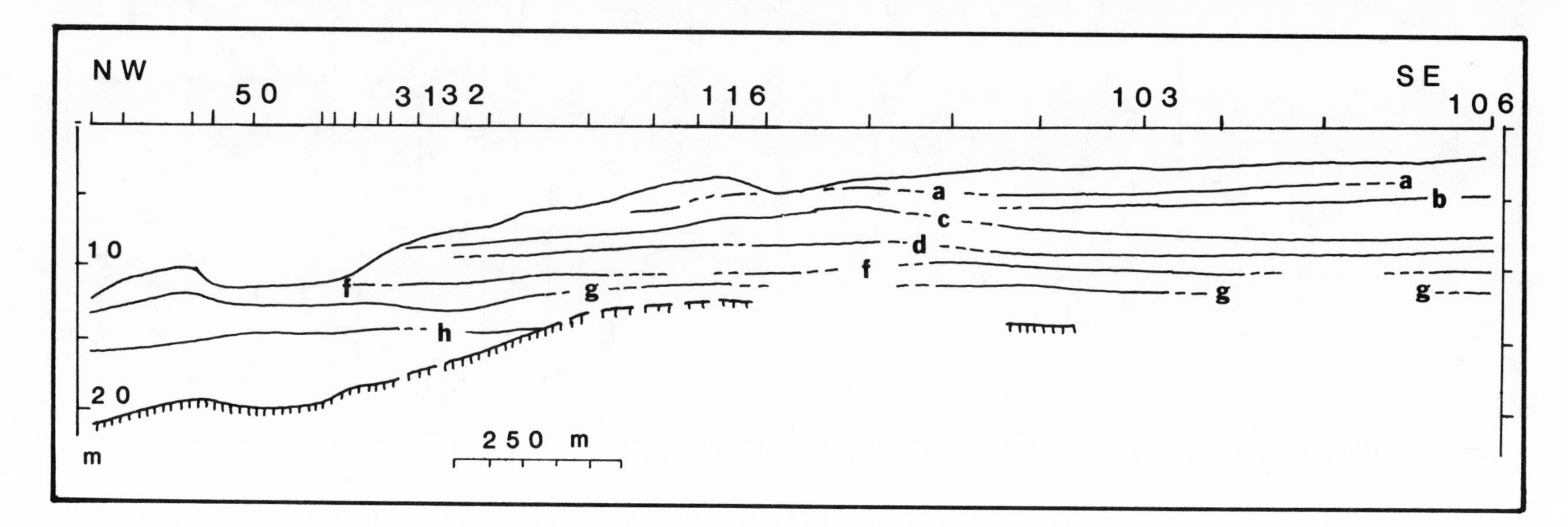

Figure 20. Longitudinal subsurface profile of Kiladha Bay along dashed line of Figure 18. The profile was constructed by connecting basal (*hachured*) and sub-bottom reflectors and the bottom at the deepest points of all traverses in Figure 18 that cross the profile line. Various stages of flooding and sediment fill are indicated by labeled reflectors. Ticks at top mark seismic profiles crossings; those shown in Figure 19 are numbered. Depths are below present sea level.

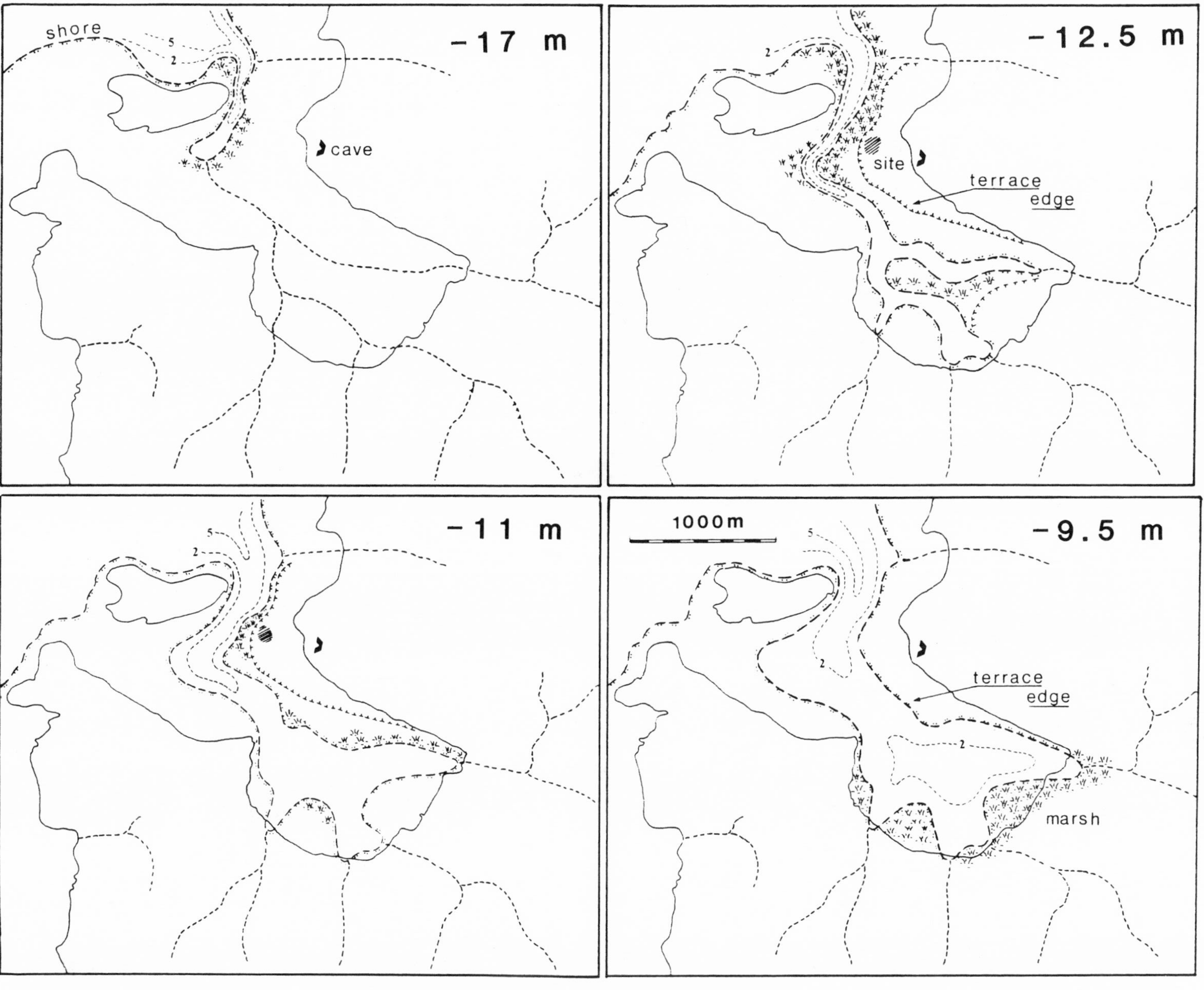

Figure 21. Kiladha Bay during the early and middle Holocene. Streams shown with dashed lines, other symbols labeled on the maps. *Upper left:* situation around 7,500 B.P. *Upper right:* the now submerged Neolithic site (Gifford 1983), possibly continuous with the onshore remains, can be seen located on the edge of a terrace overlooking a marsh and inlet below. *Lower right:* situation when sea level was at ca. −10 m, perhaps around 5,000 B.P.

the late Quaternary Mediterranean whose waters did not escape major changes anymore than did its coasts and vegetation.

The Mediterranean, an almost completely enclosed sea, segmented in several sub-basins with limited communication, differs significantly from the open ocean (Dietrich et al. 1975; Miller et al. 1970). Because the annual evaporation greatly exceeds the combined influx of freshwater from rivers and the rain, Mediterranean surface waters (Figure 22) are more saline ($37\,^{\circ}/_{\circ\circ}$ to $39\,^{\circ}/_{\circ\circ}$) than the North Atlantic (about $36\,^{\circ}/_{\circ\circ}$). Winter cooling further increases the density of the waters, enabling them to sink and form an intermediate water mass with temperatures ranging down to 12 °C and a salinity of about $38\,^{\circ}/_{\circ\circ}$. As a result, the stratification of the Mediterranean is very stable except in a few areas such as the central Aegean and northeastern Levantine basin where overturn seems marginally possible.

Not surprisingly, the regeneration of nutrients from deeper water is inhibited and especially the eastern Mediterranean is not a fertile ocean. Upwelling plays only a minor role, and the primary nutrient sources are the low salinity waters that flow from the Black Sea through the Bosporus and Dardanelles and the freshwater of the Nile. The influx from the latter has, however, been sharply reduced since the Aswan High Dam was built. Most other rivers are too small to have more than local impact. In contrast, the semi-isolated Adriatic has a higher biological productivity because of the relatively large supply of river water relative to the volume of the basin. Thus the Adriatic, the northeastern Aegean, and the southern Levantine basin are the most fertile parts of the eastern Mediterranean.

In the western Mediterranean, Atlantic surface water enters through the Straits of Gibraltar, reducing the salinity and enhancing the fertility. At a depth of a few hundred meters (the sill is slightly deeper than 300 m), this inflow is balanced by an outflow into the central Atlantic of cooler, highly saline Mediterranean intermediate water. The resulting vertical circulation ensures that, at least at the present time, even the deeper parts of the Mediterranean never become stagnant.

Given the low productivity, Mediterranean fish catches are small in terms of weight or volume, even though an exceptionally large number of species is suitable for consumption (Rand McNally 1977). It is noteworthy that shellfish and crustaceans make up about 10% of the Mediterranean catch, a much larger share than almost anywhere else in the world. Persistent overfishing for perhaps millennia as well as severe and still increasing pollution have further decreased the yield. The local catch in the Aegean region of less than 20 metric tons/year/km of coast is much smaller than that of the Adriatic or the western Mediterranean. In fact, the entire annual Mediterranean catch of 1.2 million metric tons is quite insignificant compared to the three million for just the most important fish species in the much smaller North Sea alone.

Mediterranean temperature and salinity patterns for the last glacial maximum have been determined by Thiede (1978, 1980) and Thunell (1979). In the Aegean, cores of sufficient length are available only for the southern part, but the patterns can be extrapolated northward with some confidence. Surface temperatures differed by 4° to 6° C from present mean values (cf. Figures 22 and 23), and the Aegean must have been somewhat like the present North Sea, although probably less stormy. The much lower surface temperatures reduced evaporation and thereby the amount of moisture available to locally generated storms. Diminished local storm frequencies and lower precipitation may have been partly offset, however, by a greater incidence of western cyclonic storms resulting from the more southerly position of the polar jet. The lower sea temperatures do not, therefore, explain by themselves the greater dryness of the Aegean climate deduced from the pollen record (van Zeist and Bottema 1982).

The salinity of the glacial Aegean was, at $33\,^{\circ}/_{\circ\circ}$, well below that of present and past oceanic values. This is mainly to be attributed to the reduced evaporation of a cold sea;

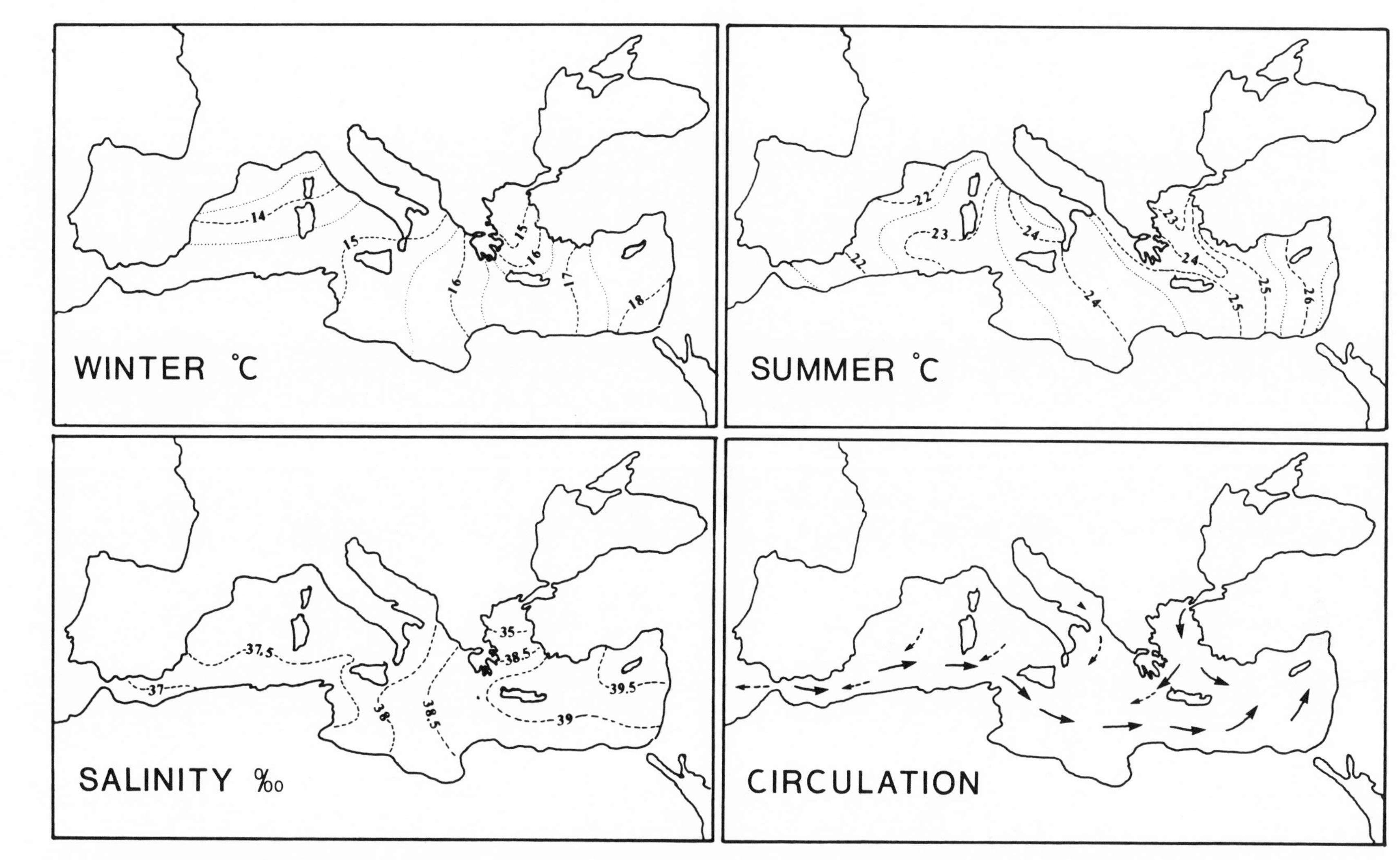

Figure 22. Winter and summer sea-surface temperatures and salinity of the Mediterranean Sea. *Lower right:* surface (*solid arrows*) and deep (*dashed arrows*) circulation. Data from Dietrich et al. (1975), Miller et al. (1970), and Rand McNally (1977).

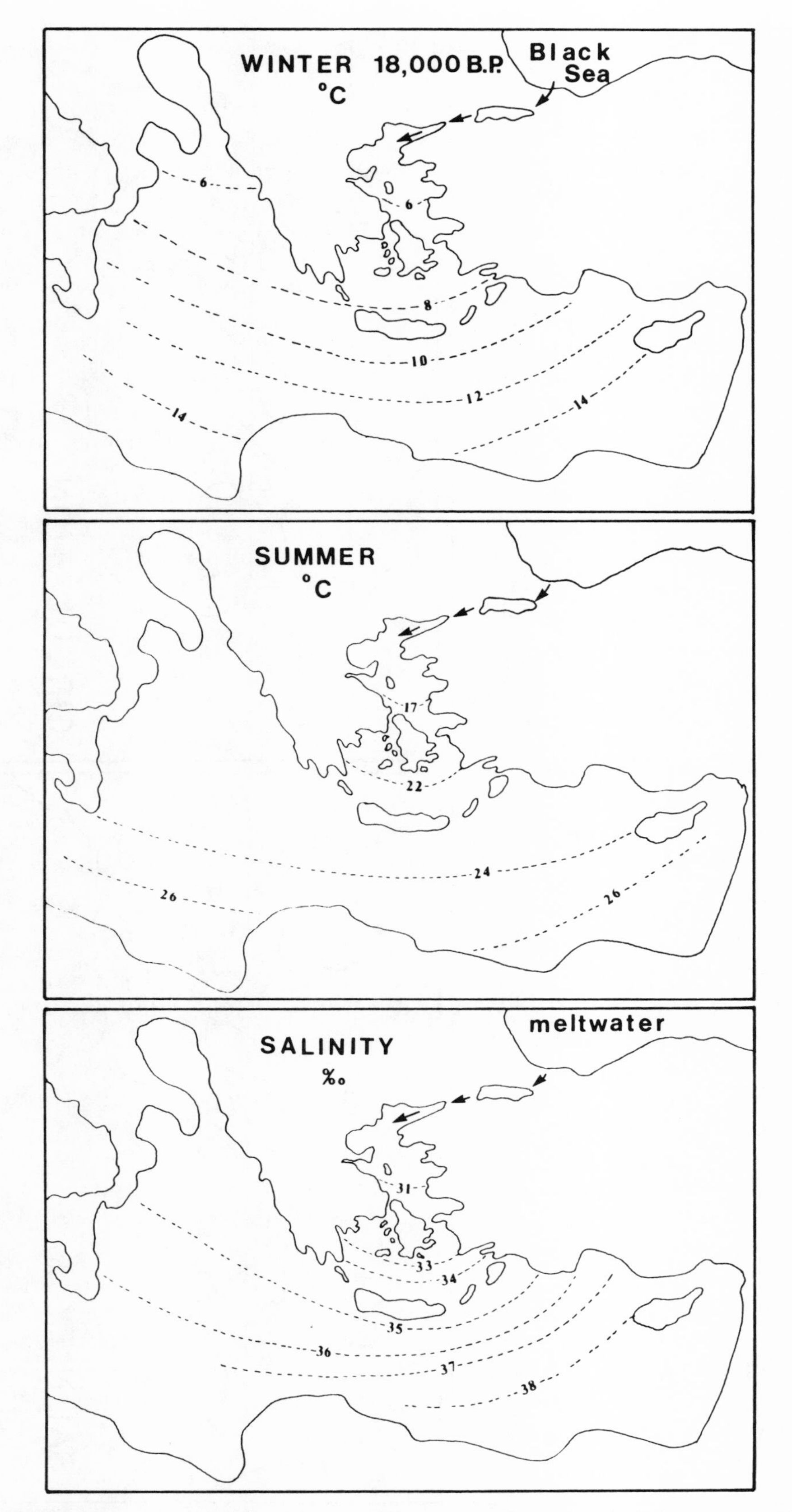

Figure 23. Winter and summer sea-surface temperatures and salinity of the eastern Mediterranean during the last glacial maximum, based mainly on Thunell (1979). Coasts are from van Andel and J. C. Shackleton (1982). Note the very reduced size and severe constriction of the Aegean Sea and the considerable cooling, salinity reduction, and greater seasonality compared with the present (Figure 22).

the influx of freshwater from melting mountain snows and glaciers may have played a minor part. Fresh or brackish water also must have entered through the Bosporus and Dardanelles, although no trace has been found (Stanley and Blanpied 1980) of the deep channels that should have connected the Black Sea with the then much lower Aegean.

The low salinity of the late Pleistocene surface waters would have reduced the stability of the water column, were it not that it was offset by the density increase resulting from the much lower temperature. On the other hand, the much altered geography of the Aegean shores, its shores everywhere close to steep continental slopes (van Andel and J. C. Shackleton 1982), would have been conducive to increased wind-driven upwelling. Moreover, the probably appreciable nutrient content of the meltwaters (Thunell and Williams, 1982, 1983) might also have led to locally increased fertility. All in all, the late Pleistocene fertility of the Aegean may have been somewhat greater than it is today, especially during the early deglaciation.

Oxygen isotope data from deep-water sediments in the southern Aegean and eastern Mediterranean show a gradual change towards present conditions beginning before 12,000 B.P. This transition was interrupted by the deposition of dark, very organic-rich sediments called sapropels which have been a recurrent feature of Mediterranean deep-water sediments since the late Pliocene (e.g., Kidd et al. 1978; Williams and Thunell 1979). The sapropels, found only in depths greater than a few hundred meters, twice appeared quite suddenly in the Aegean (e.g., Thunell et al. 1977; Williams and Thunell 1979) and in the Ionian Sea (Stanley 1978), first around 11,000 B.P., again between 9,000 and 8,000 B.P. (Rossignol-Strick et al. 1982).

The currently preferred explanation for the deposition of the Mediterranean sapropels is that they resulted from an increased stability of the water column produced by a large influx of cold meltwater from the Russian and Alpine ice caps. This would have caused near-stagnation of the bottom water, and hence a low oxygen content and good preservation of organic matter. Whether the primary productivity increased also is a matter of debate (e.g., Thunell and Williams 1982, 1983).

For the later sapropels, meltwater influx is an unlikely cause because well before their time (ca. 9,000-8,000 B.P.) the drainage from the remaining northern European ice caps had been channeled towards the North Sea (Grosswald 1980; see also Thunell and Williams 1983). Rossignol-Strick et al. (1982) hold the Nile responsible, swollen by the enhanced monsoon conditions postulated for that time by Kutzbach (1983; Street-Perrott and Roberts 1983), but that would not explain the occurrence of young sapropels in the northern Aegean (Cramp et al. 1984) and Ionian Sea.

The very limited evidence hardly suffices for inferences regarding the fertility of the Franchthi seas during the late Pleistocene and early Holocene. It seems likely that during the high glacial the fertility, especially along coasts downwind from the dominant and then probably stronger westerlies, might have been higher than it is now (Figure 24). Whether this was also true during the period between 9,000 and 7,000 B.P., when the exploitation of marine resources at Franchthi appears to have become more important (Jacobsen 1976), is not certain. The appearance of a second sapropel phase at this time suggests a greater fertility but other, and on balance perhaps more probable, causes for the preservation of organic-rich sediments may be invoked. For the greater orientation towards the sea at Franchthi at this time, one therefore leans towards other than the strictly environmental explanation preferred by Dennell (1983).

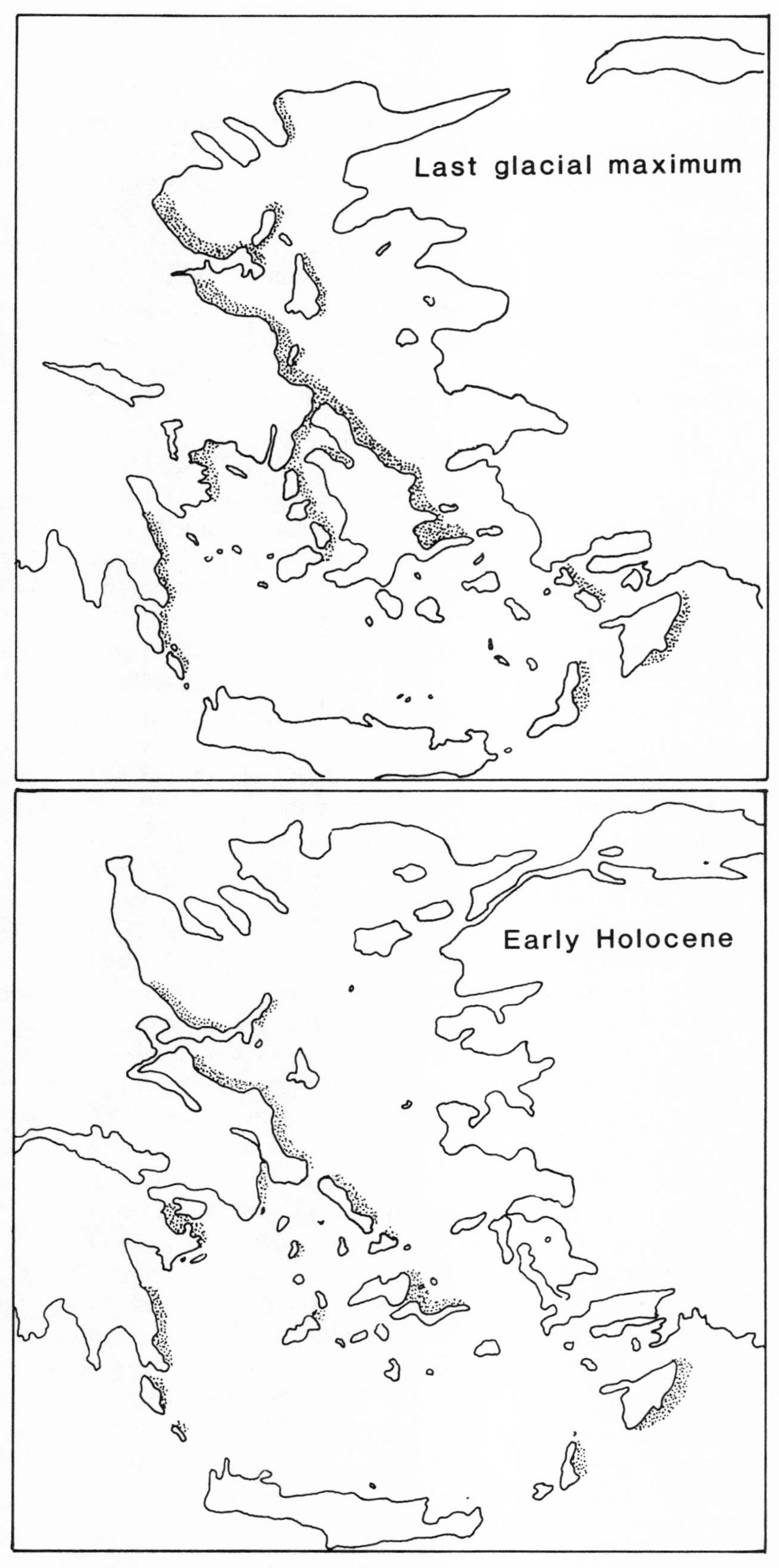

Figure 24. Paleogeography and upwelling in the Aegean during the last glacial maximum and in the early Holocene (around 9,000 B.P). Stippled areas indicate where wind-driven upwelling is likely to have occurred with westerly winds in spring. Those would be the principal areas of fertility and fish schooling, except for species which school to spawn in shallow bays during late summer through winter.

CHAPTER FOUR

Evolution of the Franchthi Landscape

Tj. H. van Andel and J. M. Hansen

The previous chapters have made it clear that the environments of the past in the Franchthi region cannot be inferred simply from its present condition. Large variations in climate and vegetation, major changes in sea level, a complex history of soil erosion and alluviation, and later strong human interaction with the land have contributed to an eventful late Quaternary landscape history. The data available for a reconstruction of this history are still less than adequate, especially with respect to climate and vegetation, but in this chapter we shall nevertheless attempt such a reconstruction as a summary and conclusion.

We may begin with what is known or may be inferred regarding the history of vegetation and climate. Our concern here is with the last glacial period and with the following early and middle Holocene warm time. During the last two million years, the northern hemisphere has passed through at least twenty cold phases of varying length and severity, alternating with shorter warm intervals (Bowen 1978; Nilsson 1983), a history considerably more complex than that of the traditional four glacials separated by interglacials.

Climate leaves only an indirect record. Although, for example, the summer and winter temperatures of the ocean surface may be derived from fossil oceanic plankton (Imbrie and Kipp 1971), or pollen diagrams may document the responses of the vegetation to changes in climate (Figure 25), the climate itself cannot usually be inferred unambiguously. Moreover, much late Pleistocene history took place beyond the range of reliable radiocarbon dates, and the precise chronology of the events, now thought to be driven in large part by the periodicities of orbital parameters of the earth (see Berger et al. 1984), continues to be debated (Kominz et al. 1979; Kukla and Briskin 1983; Morley and Hays 1981; Woillard and Mook 1982). For our present purpose these uncertainties are fortunately not critical.

Between 125,000 and 110,000 years ago, the seas were at least as warm as today (Thunell and Williams 1982), and the mid-latitude continents bore a temperate deciduous forest indicative of a climate comparable to that of the Holocene warm period. There followed a long interval of brief cold spells alternating with returns to a warmer climate (Figure 25). At about 75,000 B.P., a sharp drop of the temperature, accompanied by the virtual disappearance of deciduous forests from northwestern and central Europe and also from northern Greece (Wijmstra 1969) initiated true glacial conditions. Temporary ameliorations occurred around 60,000, just after 50,000, and before 30,000 years ago, but at no time was the improvement sufficient for a return of the deciduous forest.

Immediately after that last milder period came the coldest phase of the entire last glacial, to last for more than 10,000 years. A slight improvement began around 15,000 B.P. or so, but the true deglaciation came considerably later at a time (between 14,000 and 12,000 B.P.)

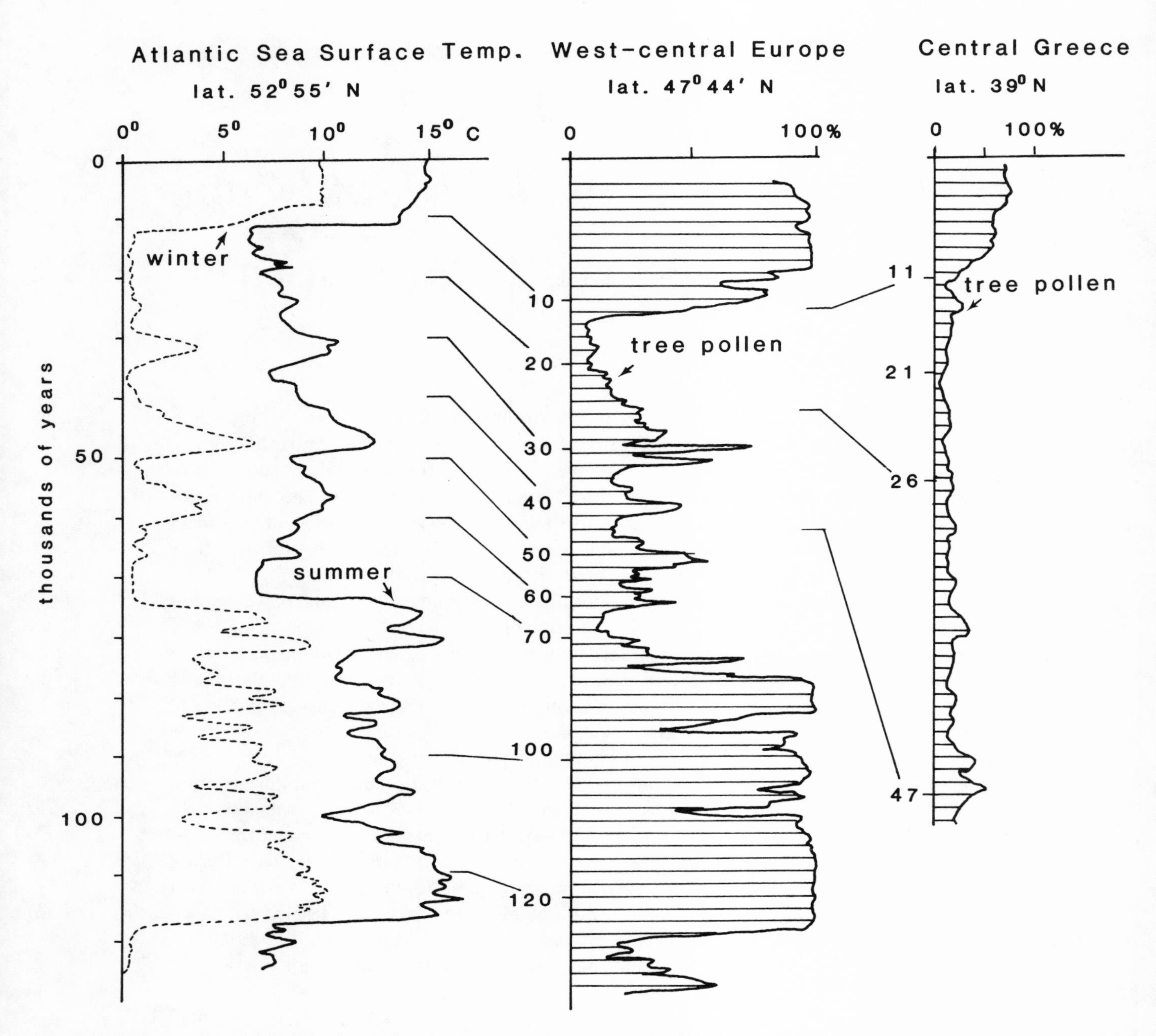

Figure 25. Late Quaternary climatic history of Europe. North Atlantic ocean-surface temperatures after Sancetta et al. (1973), west-central European pollen record from Grande Pile (Vosges: Woillard and Mook 1982), and Greek pollen record from central Greece (Lake Xinias: Bottema 1979). Chronology generalized after Kornincz et al. (1979), Morley and Hays (1981), and Woillard and Mook (1982).

which is still being debated. A rapid amelioration followed, and the postglacial climatic optimum arrived about 7,000 B.P.

The vegetation history of the eastern Mediterranean including Greece has recently been summarized by van Zeist and Bottema (1982), but their extrapolations to the Peloponnese are tenuous because of the lack there of pollen records older than ca. 4,500 B.P. During the height of the last glacial, northern and central Greece, and presumably the Peloponnese (van Zeist and Bottema 1982: Figure 14.10), were covered with a sagebrush steppe dominated by *Artemisia* sp. and *Chenopodiaceae* with some pockets of woodland. Van Zeist and Bottema speak of a "forest-steppe," but the pollen record nearest the southern Argolid (Lake Xinias: Bottema 1979) does not have a significant proportion of tree pollen. The persistence of small quantities of deciduous tree pollen, however, does imply that patches of woodland may have continued to exist in favored spots. Because the limiting factor was probably moisture rather than a low temperature, the estimated reduction of the mean annual temperature being only 2° to 5°C (Peterson et al. 1979), sheltered areas with spring-fed streams may have served as refuges. In the southern Argolid, the Iliokastro plateau may have been one such area.

Steppe or forest-steppe was still present in central and northern Greece around 12,000-11,000 B.P. (van Zeist and Bottema 1982: Figure 14.11), but when the climate began to improve, forests, first of pine, then of deciduous oak, spread rapidly. This implies an increase in precipitation as well as a rise in temperature, because otherwise the greater evaporation would have meant greater aridity. Soon after 10,000 B.P., a relatively dry oak forest with junipers (*Juniperus* sp.) and pistacio (*Pistacia* sp.) covered the lower and middle elevations of central Greece and extended into the Peloponnese as well. As the temperature and precipitation increased further, probably to a level slightly warmer and moister than today, species of a richer woodland, such as beech (*Fagus* sp.), hornbeam (*Carpinus* sp.), and hop hornbeam (*Ostrya* sp.) also became common. Van Zeist and Bottema (1982: Figure 14.12) extend this forest type across the Peloponnese, but the fact that even today the southern Argolid has the driest climate of Greece causes us to think that an open parkland is more likely than a true rich oak forest.

At this point in time, the impact of agriculture was beginning to be felt in Thessaly and Boeotia. It shows itself in the pollen records as a decline of tree pollen and an increase in the herbaceous species of wood edges and fields, many of them now common weeds. Evergreen oak (*Quercus coccifera*) appears for the first time, encouraged probably by the browsing of domesticated flocks. Known to visitors of Greece mainly as a prickly, multi-trunked shrub, it does grow into a substantial tree if left alone (Plate 1b), and evergreen oak woodland can still be found here and there in Greece (Rackham 1983).

Around 5,000 B.P., the vegetation histories of various parts of Greece began to diverge in response to differences in the human impact on the landscape. The pollen record is therefore no longer a useful climatic indicator. Around Lake Kopais in Boeotia, for example, large-scale clearing of woodland took place about 5,000 years ago (Turner and Greig 1975), while in Macedonia major deforestation did not begin until well into historical time (Wijmstra 1969). Contrary to widely held belief, the greater part of Greek forests was lost even more recently, largely in the nineteenth century (Meiggs 1982).

For the vegetation history of the southern Argolid we are limited to Hansen's (1980; forthcoming) study of plant remains from the Franchthi excavation which covers the period of habitation of the cave until ca. 5,000 B.P., and to pollen diagrams from coastal lagoons for the later period (Sheehan 1979). Both have limits for the study of the natural vegetation of the area. The cave deposits contain mostly plant remains brought intentionally or accidentally by the inhabitants, probably mainly from the vicinity of the cave. The pollen data reflect a wider area, but lagoonal sediments often contain hidden hiatuses (Wright 1972), and the number of radiocarbon dates is small.

The plants from Franchthi Cave suggest an open sagebrush steppe for the late glacial, in accord with the animal bone remains (Payne 1975, 1982), and confirm a considerably cooler and drier climate. If deciduous trees had found a shelter somewhere, the cave deposits provide no evidence for it.

As elsewhere, the vegetation became richer around 11,000 years ago. The cave deposits of that time contain wild almond (*Prunus amygdalus*) and pear (*Pyrus amygdaliformis*), bitter vetch (*Vicia ervilia*) and wild lentils (*Lens* sp.), and wild oats (*Avena* sp.) and barley (*Hordeum* sp.), all potentially found on the valley floors and the slopes of the southern hills, though surely not on the rugged limestone ranges. Pistacio (*Pistacia* cf. *lentiscus*) may have come from stream banks, and various grasses and bulbs from low ground in what is now Kiladha Bay. The streams were deeply incised at the time, and flood plain resources must have been small.

We interpret this to mean that perhaps as early as 10,000 years ago the area had become covered with an open wood or parkland, probably of oak and/or pine, although the cave deposits have so far yielded no confirmation of the presence of either. An understory of wild pear and pistachio would have been likely. Some steppe may have remained for a while on the coastal plain, dwindling rapidly as the sea rose. A climate much moister than before is evident, although it may have remained relatively cool for another few thousand years.

The flora continued to diversify until 7,000 or 8,000 years ago, and the deep soils of the rolling hills in the south and on the deeply weathered ophiolites of the Fourni valley became covered with an open wood, probably of deciduous oak, beech, holly, hornbeam, and other species no longer found in the area. Again we have little direct evidence for the presence of various trees in the area at that time other than those that produced edible fruit, but some of them were found in the oldest pollen spectra (ca. 4,500 B.P.) of Thermisi lagoon (Sheehan 1979). The streams continued to be edged with wild pear, pistachio, and probably plane trees and maples, because the groundwater level had not yet been lowered by human overuse. On the other hand, the rugged limestone ridges lacked even then soils deep enough to support a significant tree cover (Jameson et al. forthcoming: Chapter 3.2) and must have carried some form of maquis.

Our knowledge of the vegetation history after ca. 5,000 B.P. rests entirely on Sheehan's (1979) pollen data. The earliest spectra, dated at 4,400 B.P. in uncalibrated radiocarbon years, still contain evidence for a deciduous oak forest with holly (*Ilex* sp.) and horse chestnut (*Aesculus* sp.), but about a thousand years later, maquis and pine woods, with intermittent evidence for olive culture, dominate the vegetation history.

Turning from vegetation and climate to other aspects of the history of the Franchthi landscape, we note that during the entire known occupation of the cave the landscape appears to have been geomorphologically stable, at least in general. The upper Loutro alluvium, deposited during a mild interstadial between 45,000 and 30,000 B.P., precedes the occupation and the Pikrodhafni, formed 4,500 years ago, comes afterwards. During this long period of time the streams were incised, the slopes stable, and we have no evidence for any impact of human use of the land on erosion and sedimentation. This suggests either that land clearing for farming was not yet extensive, even during the later Neolithic, or that it was accompanied by very long fallow. Farming may also have been restricted to less vulnerable land such as the spring-watered terraces along the shores of Kiladha Bay.

Far from minor, on the other hand, was the last factor that we must consider, the large change in land area that resulted from the fall and rise of the sea in response to continental glaciations and deglaciations (Table 3). We have discussed in Chapter 3 the details of

TABLE 3

LATE QUATERNARY CHANGES IN COASTAL-PLAIN AREA IN THE SOUTHERN ARGOLID

Period	*Hill/mountain area*	*Coastal-plain area*	*Coastal plain in % of total area*	*Coastal plain in % of area at 20,000 B.P.*
At present	372.9 km^2	36.8 km^2	9.0%	15.0%
2,500 B.P.	372.9	47.7	11.3	19.4
5,000 B.P.	372.9	57.3	13.3	23.3
8,000 B.P.	372.9	79.8	17.8	32.5
10,000 B.P.[a]	400.3	112.3	21.9	45.7
20,000 B.P.[b]	407.9	245.4	37.6	100.0
35,000 B.P.[a]	400.3	113.8	22.1	46.3

[a]includes Dhokhos and Spetsai islands as hill/mountain area

[b]includes parts of Idhra and associated islets as hill/mountain area

N.B.: The area considered is that shown on Figure 27, south of the Dhidhima and Adheres ranges.

---

changing coastal environments in the Franchthi embayment; here we summarize the broader changes for the entire peninsula. Seismic reflection traverses (van Andel and Lianos 1983) allow us to do this with considerable precision for the southern Argolid, and bathymetric data, requiring little adjustment for the thin postglacial deposits, extend the positions of past coastlines around the entire Argolid (van Andel and J. C. Shackleton 1982).

Prior to the last glacial maximum, our best estimate suggests a sea level 30 m or 40 m below the present one (Figure 9). Consequently, a coastal plain from several hundred meters to a few kilometers wide surrounded the southern Argolid (Figure 26). It included the islands of Spetsai and Dhokos and was probably covered with sagebrush steppe and patches of woodland. It would have provided grazing for sizable herds of herbivores.

After 30,000 B.P. the sea sank to ca. −120 m, and the coastal plain greatly increased in size. For some 10,000 years it nearly equalled the present land area of the southern Argolid, adding a vast level plain to its rugged hills and mountains. The island of Idhra and its many islets were attached to the mainland, forming prominent hills in the wide steppe. This coastal plain extended from the head of the Gulf of Argos around the tip of the peninsula to Methana and thence to Attica, enclosing a large lake in the western Saronic Gulf (Figure 27).

The evidence favors a cold, dry climate, and the plain was surely covered with steppe and with few trees, but it was far from waterless. The streams, though larger than their present scanty remnants, must have been ephemeral, but many of the springs that are now below sea level (Figure 5) should have continued to flow. Endowed with so many oases and watering holes, the plain should have been able to feed large herds of such animals as *Equus hydruntinus* and, if it was sufficiently moist in certain places or at certain times, bovids such as the aurochs (Payne in Herz and Vitaliano 1983).

When finally the sea began to rise, the coastal plain shrank rapidly. Idhra and adjacent islets became detached first, the land bridge having been a mere 20 m above lowest sea

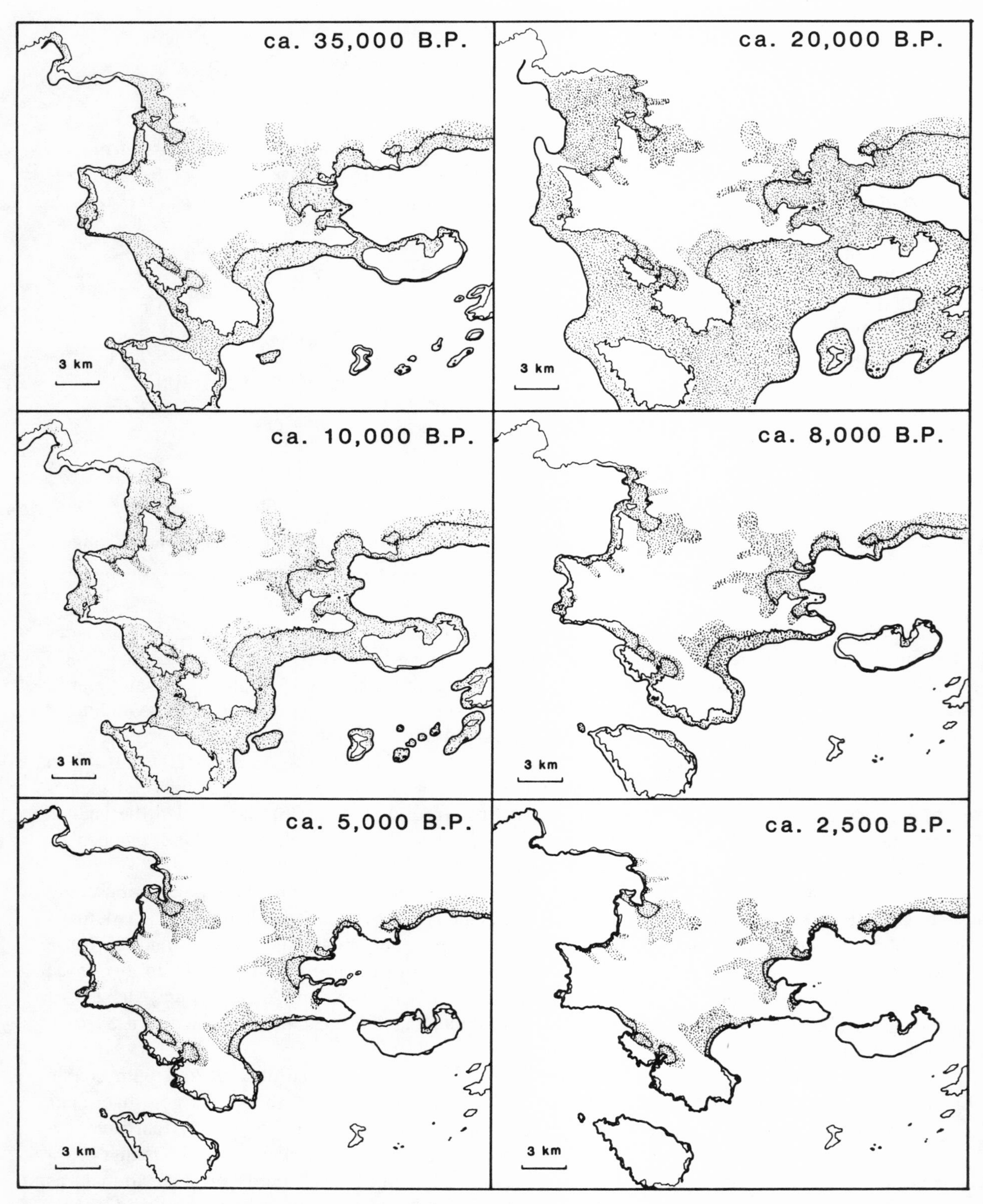

Figure 26. Paleogeography of the coast of the southern Argolid in the late Quaternary. The 35,000 B.P. shore is based on an assumed sea level of 30-40 m below the present one. The others are located at the lowest level in the region (−120 m), at −54 m (ca. 10,000 B.P.), −29 m (ca. 8,000 B.P.), −10 m (ca. 5,000 B.P.), and −6 m (2,500 B.P.). Data from van Andel and Lianos (1983). Coastal plains and lower valleys have been stippled.

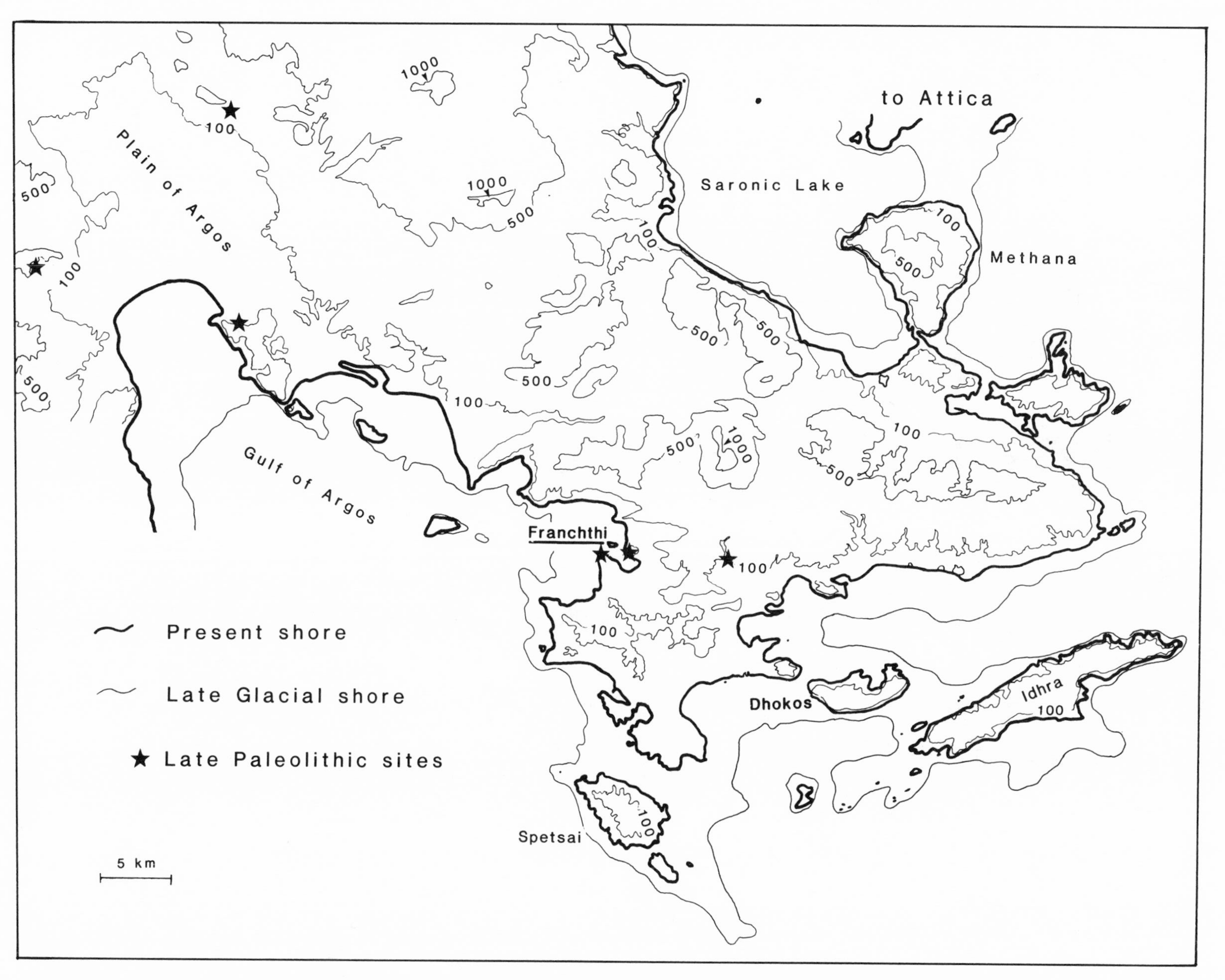

Figure 27. The Argolid peninsula during the last glacial maximum. Contours in meters after Greek topographic maps, scale 1:50,000. Late Paleolithic sites along Argive plain are, from left to right, Kefalari Cave, Klisoura Caves, and cave at Navplion.

level. Dhokos and Spetsai reverted to island status about 9,000 years ago, but a slowly narrowing strip of coastal plain continued to surround the southern Argolid for several thousand years more. The coast, today mainly rocky or cobbly and set against steep bluffs, was for a long time fringed with sandy beaches and marshes.

The gradual loss of level coastal land is one of the largest environmental changes of the entire late Quaternary. The loss continues today, though slowly, and only a little outbuilding of the coast has occurred because the sediment supply is so small. Archaeologically, the loss of coastal plains is arguably the most important postglacial environmental event, in the beginning because of the diminishing resources available to the hunter-gatherers, and later as it reduced the most valuable arable land, especially the fields near coastal springs.

PART II

The People

Susan B. Sutton

CHAPTER FIVE

The Franchthi Region in Modern Times

The harvesting rigs that now lumber along the paved roads of the southern Argolid and the tourist hotels that fill and empty to the rhythms of the hydrofoil from Athens seem to have little in common with the Paleolithic and Neolithic periods discussed elsewhere in this series of publications. That the eastern portion of the Franchthi headland has recently been removed by lime quarrying heightens this sense of discontinuity between past and present. This chapter aims to demonstrate, however, that, while some current activities are indeed qualitatively different from those of the past, the abstracted economic and settlement patterns that emerge from these activities have clear and significant ancient parallels.

This understanding is achieved only partially by looking at technological practices of great antiquity. Equally important is consideration of the southern Argolid's role in the global, industrialized systems of modern times and the impact these have had on its internal organization. Such analysis reveals certain economic and demographic patterns resembling those of particular periods of the area's past. As van Andel points out earlier in this volume, the people of this region have chosen parallel solutions at different times to meet similar problems. The constant in this equation is the geographical location. The overarching variables are the economic and political systems in which this location participates. Certain systems create certain patterns of local response whether they occurred four thousand years ago or today. The modern period involves a transition from a time of relative isolation to one of strong involvement in external systems. This chapter considers how the economic strategies and settlement patterns of the southern Argolid have recently changed in response to this transition.

The modern period, which stretches from just before the Greek Revolution to the present, forms a unified historical unit. During this time Greek lands moved from a decentralized, feudal system to a capitalized one, strongly directed by the national government and industrialized powers abroad. The region around Franchthi Cave played an interesting role in these developments. Indeed, the modern history of the southern Argolid[1] stands in some contrast to many other areas of Greece. While depopulation and economic failures have characterized much of provincial Greece, the southern Argolid has remained demographically stable and economically viable. Understanding why this has been so not only provides ethnoarchaeological parallels with previous periods when the region occupied a similar position, but thus also illuminates one of the major issues of modern Greek life.

The regional social history necessary for this analysis is made possible by the considerable literature that exists on the southern Argolid. This region and nearby areas have attracted an astonishing number of anthropological studies, including those by H. Koster, J. Koster, and Chang on pastoralists, those by H. Forbes, M. Clark Forbes, Gavrielides, Murray, and Kardulias on agriculturalists, Bintliff's research on local fishermen, and my own on regional population dynamics.[2] Karanikolas and Petronoti also have recently published historical studies of the area's shipping/merchant class. Topping's analysis of the southern Argolid in the century prior to independence provides additional valuable background. Finally, the area was significant enough to be mentioned in most of the major histories of modern Greece and to have received visits by many travellers and scholars during the last two centuries.

This relative abundance of information is here woven together to correlate the southern Argolid's changing economic and settlement patterns with its involvement in the increasingly centralized systems of modern Greece. The area's economic history is first presented in several chronological phases from the eighteenth century to the present. The impact of this history on modern settlement patterns and the importance of the modern period for studies of the region's past follow. Throughout this presentation, although the entire southern Argolid is considered as a functional regional unit, the emphasis is on those areas closest to the Franchthi Cave (Figure 28).

CONDITIONS IN THE EARLY EIGHTEENTH CENTURY

The regional alignments of the southern Argolid began a dramatic transformation in the late eighteenth century. The conditions in the area just prior to that time form the backdrop against which those changes must be understood. Greek lands were then largely under Turkish control, a brief Venetian challenge in the Peloponnese having been rebuffed in 1715. The Ottoman system, although overseen from Constantinople, allowed considerable local autonomy in many provinces (Vacalopoulos 1976:205). The southern Argolid was one of these. Turkish rule of this part of the Peloponnese was centered in Navplion, and the southern peninsula was largely left to its own devices. The inhabitants of Kranidhi sided with the Turks in the Venetian-Turkish War of 1715 and were rewarded afterward with both administrative concessions and land (Karanikolas 1980:35). Two large, land-owning Orthodox monasteries provided additional local power bases: Ay. Dhimitrios tou Avgou in the northwest (Plate 8b) and Ay. Anaryiroi near Ermioni (Topping 1976).

The population was clustered into four villages, each located in one of the southern Argolid's four arable zones (Bintliff 1977; van Andel et al. 1986) (cf. Figure 28). These areas had been farmed, abandoned, and then farmed again for millennia, something evident both from archaeological finds and soil erosion patterns (van Andel et al. 1986). Topping's analysis (1976) of the early eighteenth century cadaster produced during the brief Venetian reign in the area gives both population size and economic base for these four villages. Dhidhima was the smallest with between 150 and 200 residents, then Fourni with around 260 inhabitants, Ermioni with 375, and finally Kranidhi with approximately 1,000. A few small hamlets, isolated farmsteads, and seasonal shepherd huts also existed (e.g., Gell 1810; Houliarakis 1973:1:40; Pouqueville 1826:5:257). Many of the inhabitants were Albanian speakers, descendants of several waves of Albanian migrants who entered the area in the fourteenth to sixteenth centuries (Gavrielides 1976b:143; Karanikolas 1980:17; Paraskevopoulos 1895:57).

The two smallest settlements were based on farming and shepherding. Large numbers of sheep and goats were recorded as the major wealth of Dhidhima (Topping 1976:98). Fourni is listed with both herds and olives, something consonant with Gavrielides' belief (1976b:157) that the many olives cultivated there during Byzantine times had largely been abandoned and that the valley was mainly used for shepherding when reoccupied in the

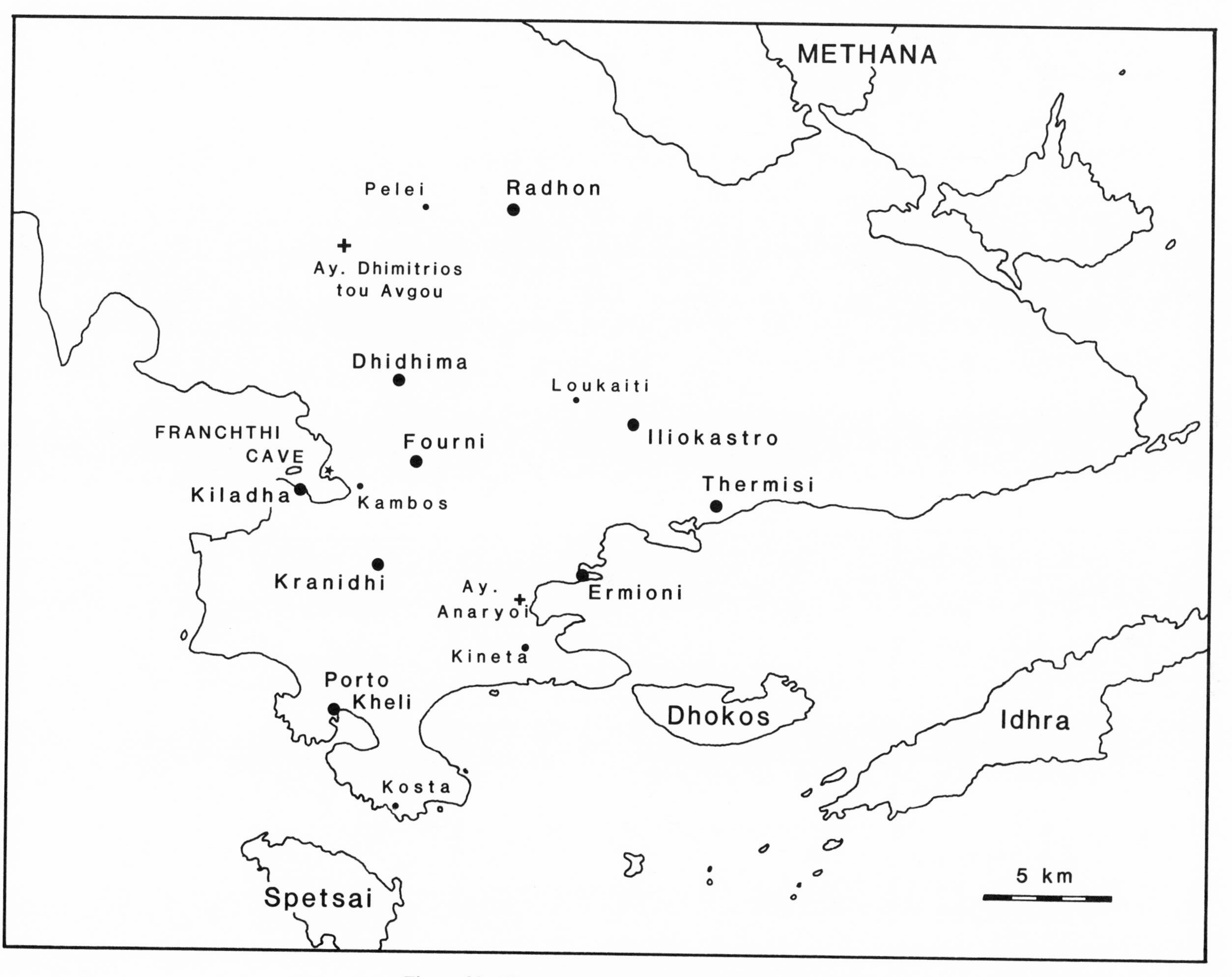

Figure 28. Place names in the southern Argolid.

fourteenth and sixteenth centuries. Both olives and carobs are recorded as the produce of Ermioni and Kranidhi (Topping 1976:100). It seems likely, however, that maritime activities were also pursued in these two larger settlements. Ermioni's location on the sea and local records showing at least one corsair-like family had moved into Kranidhi by this time (Karanikolas 1980) indicate this. Whatever had already developed along these lines, however, was only a precursor of what was to occur in the next phase.

Since export agriculture was not very important to the southern Argolid at this time, farming probably involved many of the subsistence techniques still used. H. Forbes' discussion (1976) of certain farming practices on nearby Methana seems especially relevant. Polycropping and holding small plots scattered through several ecological zones stabilize a subsistence economy but become less functional in those oriented toward cash production. Koster and Koster's analysis (1976) of the symbiotic relationships that can exist between shepherding and farming may also have held true. Both activities are recorded for the southern Argolid at this time. Sheep and goats can be pastured in marginal or fallow land without hindering farming, and produce is frequently exchanged between shepherds and farmers. More than one set of pastoralists may have been using the southern Argolid. Three currently do: the local Albanian-speakers of Dhidhima, the transhumant Valtetsiotes who winter on the Thermisi plain, and the Sarakatsani who winter somewhat further north. M. Clark Forbes' study (1976a) of the gathering strategies which have long accompanied shepherding and farming can also be carried back in time. Gathering wild greens, mushrooms, berries, bulbs, nuts, herbs, snails, and shellfish are activities that can fit with the agricultural work cycle. The production of firewood, charcoal, lime, and resin from the natural environment (Forbes and Koster 1976:121; Gaverielides 1976b:143) may also have been carried out.

The immediate vicinity of Franchthi Cave had little settlement at this time. The prehistoric coastal plains in front of the cave were submerged and no longer available for agriculture (van Andel, this volume). The closest sizeable village was Fourni, although there was also a small cluster of buildings on the southern side of Kiladha bay. The small monastery of Zoodokhos Piyi had been built there in 1527 and may have had a few houses attached to it (Pouqueville 1826:5:261). Judging by its nineteenth century use, the bay probably harbored small boats from Kranidhi.

EXPANSION AND REVOLUTION

The relative freedom that the Ottomans had granted the southern Argolid by the early eighteenth century led the area to occupy a position of some importance by the time of the Greek Revolution. During this period the feudal systems of the Ottoman Empire were coming more and more into contact with the expanding industrial, capitalist states of western Europe (Mouzelis 1978). Certain categories of Greeks played pivotal roles in mediating the two economies, and some of these soon also took up the idea of Greek independence. Although outsiders and bureaucrats now sometimes view the southern Argolid as a backwater, it clearly belonged in this small group of dynamic regions directly responsible for bringing about the Greek Revolution in the late eighteenth and early nineteenth centuries. The transformations undergone by the southern Argolid during that period and described in this section are also connected to the area's current stability.

The southern Argolid followed the lead of the nearby offshore islands of Idhra and Spetsai in these developments. The location of these islands near major sea routes, their military defensibility, and the independence conceded them by the Turks enabled both to become major sea powers during the eighteenth century. The spectacular rise of Idhra and

Spetsai from lowly populated, minor islands to the center of Greek shipping accompanied the growing desire of Russians, Greeks, and Ottomans to have their produce enter Mediterranean trade routes (Kriezis 1860:38; Michaelides 1967:12-16) and reach a growing western European market (Petronoti 1985:64). The Venetians, who had long been the primary carriers in this part of the Mediterranean, had just been ousted by the Turks. Idhra and Spetsai were granted concessions by the Sultan to step into this vacuum.

These islanders, who had begun as small-scale coastal traders, thus expanded their operations to international shipping. The French Revolution and Napoleonic Wars increased this trade as Russian and Turkish wheat was transported to blockaded ports in Europe (Kriezis 1860:38; Michaelides 1967:19; Petropulos 1968:20; Orlandos 1877:27). As a result, the populations of Idhra and Spetsai rose to as much as 40,000 and 25,000 people, respectively, by the early nineteenth century (Houliarakis 1973:1:29), and tremendous wealth was generated on them. While this led to an oligarchy on Idhra, the profits were still filtered, though in lesser fashion, to all who participated. Frequently, for example, sailors' wages represented a set share of the profits from a voyage (Finlay 1877:6:168).

Such events reverberated in the southern Argolid, the mainland closest to these islands. There were strong social ties between the southern Argolid and the islands. Before their rise to power, Idhra and Spetsai had been almost deserted. They were slowly populated during the fifteenth to sixteenth centuries by the same waves of Albanian refugees fleeing the advancing Ottoman empire as had settled the southern Argolid (Kolodny 1974:1:177; Kriezis 1860:9-17). Another group of settlers in the seventeenth century was drawn almost entirely from Ermioni and Kranidhi. Thus, as Idhra and Spetsai began their rise to power, kinship ties already connected them to the southern Argolid (Orlandos 1877:13). Connections between the areas intensified as the islands grew in power. By the late eighteenth century, the colony of Kranidhiotes seeking their fortunes on Idhra numbered at least 600 people (Kriezis 1860:17). Wealthy families and monasteries on Idhra and Spetsai also acquired agricultural estates on the mainland. The Idhriote monastic *metochi* on the Thermisi coast, the Voulgaris estate near Ermioni (Kriezis 1860:7; Paraskevopoulos 1895:52) (Plate 8a) and the Spetsiote estates near modern Kosta (Miliarakis 1886:226) helped feed the growing island populations.

Such ties led the southern Argolid to participate in the growth of Idhra and Spetsai in several ways. The population of Kranidhi went from approximately 1,000 in the early eighteenth century (Topping 1967:99) to 4,813 in 1829 (Houliarakis 1973:1:40), as people moved to be near the centers of the shipping boom. While the origins of these migrants are unclear, what they did when they got to the southern Argolid is not. Some engaged in agricultural and pastoral activities which had been stimulated to meet island markets (Jameson 1976:77; Koster and Koster 1976). Most of the area's residents, however, were directly involved in the sea trade itself. Some Kranidhiotes owned and captained ships sailing to the Black Sea, North Africa, and Asia Minor (Petronoti 1985:65-66). Many others of the region were sailors, and indeed the "boatmen of Poros, Kastri (Ermioni), and Kranidhi" are mentioned as the most significant by Finlay (1877:6:30) in writing of this time.

It is of little surprise that these rather unique places of Greek merchant capitalism gradually began to align themselves more with western Europe than with the Ottomans (Finlay 1877:6:166; Orlandos 1877:34; Petropulos 1968:35). Wealthy Idhriotes and Spetsiotes were influenced by their education toward the Hellenic rather than Romeic view of Greek destiny, a view connected to the growing classical interests of western Europe (Herzfeld 1982). The decline in grain prices in Europe after 1815 placed a strain on Mediterranean shipping that caused some to want changes in Ottoman maritime regulations (Finlay 1877:6:169). Once the Greek Revolution had begun, Idhra and Spetsai contributed 175 large ships to the

Greek navy, more than any other area of Greece. Their leaders also constituted one of the major factions directing the fighting and organizing the subsequent modern Greek state.

The southern Argolid once again followed the lead of Idhra and Spetsai in these matters. It quickly joined the Revolution and was closely allied with the islands' political faction (Antonakatou and Mavros 1973:139-145; Karanikolas 1980:19). Ermioni, Kranidhi and the monastery of Avgo were all, at various points, revolutionary centers. A provisional government for Greek forces was temporarily established under the direction of Idhriote leaders at Kranidhi (Petropulos 1968:84), and Ermioni was the seat of the Third National Congress in 1827 (Antonakatou and Mavros 1973:149).

SHIFTING FORTUNES, SHIFTING STRATEGIES

Although Idhra, Spetsai, and the southern Argolid entered into the new nation from positions of some importance, their fortunes soon took a turn for the worse. Their role in connecting feudal and capitalist structures was no longer as important, and their position within the emerging structures of the modern Greek state was less secure. The 150 years since the Greek Revolution have witnessed increasing national centralization around Athens, which not only houses the national government but is also the point of greatest importing and exporting, the financial hub, the industrial leader, and the educational center for all Greece. The centralized governmental structures imposed by the northern European guaranteeing powers (Petropulos 1968) and the characteristically underdeveloped economy resulting from the strong influence of these same nations on Greek capitalism (Mouzelis 1978) have rendered this so. Greece's internal center and major point of contact with outside systems are combined in one primary city.

Only a few other towns and rural regions have managed to maintain demographic and economic health in this situation. Most have lost much of their population through migration. This section examines the fate of the southern Argolid under these conditions in two steps. First, the nineteenth century is identified as a time during which the area retained some of its former importance. Second, the collapse of Greek shipping with the rise of steamships at the end of this century is revealed as marking a new phase. As the narrative reaches the present day, it will appear that while Idhra and Spetsai have undergone much decline, the southern Argolid has weathered these events somewhat better.

NINETEENTH CENTURY ADAPTATIONS IN MARITIME COMMERCE

Although Idhra, Spetsai, and the southern Argolid eventually lost their importance in Greek shipping, this did not take the sudden catastrophic form by which it is sometimes portrayed. The preeminent place among Greek ports moved first to Syros and then, by 1870, to Athens (Kolodny 1974:1:103-112, 192-199).[3] Idhra, Spetsai, and Kranidhi did, however, maintain lesser but still active harbors until the advent of steamships in the late nineteenth century (Karanikolas 1980:20; Orlandos 1877:39; Paraskevopoulos 1895:37-46; Petronoti 1985). Spetsai, for example, still had 190 ships in its harbor in 1860 (Orlandos 1877:39-47), and Kranidhi had 38 large ships and 80 smaller ones to its name in 1873 (Karanikolas 1980:20). Furthermore, most of the working population of the southern Argolid still actively pursued sea occupations throughout the nineteenth century. Almost three-quarters of the area's population lived in Kranidhi and its two ports, and well over half of these families were involved in maritime activities (Paraskevopoulos 1895:48). The nature of these activities gradually diversified to meet the changing economic circumstances of the region, but the importance of the sea remained constant.

Whether sailing on local, Syriote, or Athenian ships, many men from the southern Argolid continued to work in the Greek merchant marine. In the 1830s, half of the sailors in Greece came from Kranidhi, Idhra, and Spetsai (Strong 1842:158-159). Kranidhi itself was listed third in number of sailors, just behind Idhra and Syros, but above Spetsai. Kolodny (1974:1:348) concurs that men from the Argosaronic region have been disproportionately numbered among the Greek merchant marine, and he notes that, in 1879, one-fourth of Greek sailors were from that general region.

The southern Argolid also became known for ship construction during the nineteenth century. Kranidhi, especially, was known for building small, wooden ships, similar to those still sometimes assembled in Kiladha and Ermioni today (Plate 12c). Some 600 ships from the Kranidhiote shipyards of Porto Kheli and Kiladha appeared in the harbor registers of Idhra and Spetsai at the peak of this construction around 1870 (Petronoti 1985:66). In the 1830s, Kranidhi stood twelfth among all places where Greek ships were built (Strong 1842:149). Those vessels used locally transported Peloponnesian produce to Athenian markets or participated in the area's growing sponging industry.

As long-distance shipping declined, the sailors and shipowners of the southern Argolid turned to sponging (Karanikolas 1980:34; Kriezis 1860:197; Miliarakis 1886:241; Paraskevopoulos 1895:32; Petronoti 1985:72). Idhra, Spetsai, Kranidhi, Ermioni, and Fourni are all cited as sponging centers well into the twentieth century. Greek sponging began in response to increased northern European demand for both industrial and cosmetic sponges (Georgas 1937; Kolodny 1974:1:309-323). Greek sponge banks were soon depleted, and those off the north coast of Africa began to be exploited by the middle of the nineteenth century. While Kalymnos and other eastern Aegean islands not yet part of the Greek state were most active in this, the Argosaronic region was second in importance. The opening of sponge beds in the new world and the creation of synthetic sponges in the twentieth century eventually ended this trade, but for a while it proved a valuable refocusing of the area's seafaring abilities.

Agricultural production continued to be stimulated by the area's maritime activities (Gavrielides 1976c:273; Koster and Koster 1976:281; Petronoti 1985:67). Given the environment of the southern Argolid, agricultural expansion is possible there only with large inputs of either capital or labor (van Andel et al. 1986). Both were now available. As shown in the demographic section later in this chapter, the region experienced positive population growth rates at this time. The wages of sponging and sailing provided the capital. Local produce could easily be taken to expanding Athenian markets. Furthermore, maritime activities could be synchronized with the agricultural cycle (Gavrielides 1976c:273). Gavrielides believes Fourni olive production expanded beyond self-sufficiency at this time (Plate 7), and van Andel (this volume) concurs that olive cultivation increased in the southern Argolid after Greek independence. Miliarakis (1886:230) lists olives as the major export crop for the area at this time, with wine, shepherding, honey, carobs, and resin following in that order (see also Paraskevopoulos 1895:49).

THE CRISIS IN GREEK SHIPPING AND TWENTIETH CENTURY ADAPTATIONS

The invention of the steamship dealt a serious but temporary setback to Greek shipping in general, and a permanent one to that of Idhra, Spetsai, and the southern Argolid (Karanikolas 1980:20; Kolodny 1974:1:328; Orlandos 1877; Paraskevopoulos 1895:37-39; Petronoti 1985:72). The wealthy families of this region had been reluctant to invest in such ships; those in Athens had not. All Greek shipping suffered temporarily, and when Greek fleets were finally converted to steam, they were concentrated around the capital city. Idhra and Spetsai lost most of their shipping activities after 1870. The Kranidhi fleet fell from approximately 135 ships to 75 in one decade (Miliarakis 1886:240).

While this crisis greatly impoverished Idhra and Spetsai, the southern Argolid proved more resilient. The population of the two islands plummeted, but that of the southern Argolid remained relatively stable (Table 4). The economic base of Idhra and Spetsai is now almost exclusively tourism, while the inhabitants of the mainland have continued to follow a wide variety of occupations. The two areas thus progressively differentiated from each other, something given formal recognition when the Eparchy of Spetsai and Ermionis was divided into two eparchies in 1951. While connected to Athens only through the weekly passenger ship to Idhra and Spetsai in the 1870s, both Ermioni and Porto Kheli now receive several ships daily (Kolodny 1974:1:108-113). The road built between Kranidhi and Navplion in the 1950s opened the area to land connections for the first time (Jameson 1976:77). Such developments enabled the southern Argolid to become an area of moderate prosperity, well-integrated into the national system in its own right.

The economic stability of this region has rested on its more direct access to Athens than most of the Peloponnese, its greater suitability for agriculture than Idhra and Spetsai, and the reinvestment of local capital in a variety of new economic pursuits following the shipping decline. Some of this wealth belonged to the former shipowners and captains who turned to other enterprises (Petronoti 1985:72). The rest was earned through wage work in a variety of activities, but especially those connected to the sea or earned through migration. Gavrielides (1976b:143) has documented how Fourniotes combined fishing, sponging, shepherding, cash cropping, pine tapping, seafaring, and cyclical migration to provide a flexible base of operations.

THE RANGE OF CURRENT ECONOMIC ACTIVITIES

Two maritime activities are currently of greatest importance to the southern Argolid. The first continues to be the Greek merchant marine. While large transport ships no longer depart Kranidhi's harbors, men from Kranidhi and the coastal villages sign on those leaving Piraeus. The residents of Kranidhi, Ermioni, Kiladha, and Porto Kheli repeatedly mention this as the most common and significant occupation for young men. Over 1,000 sailors lived in Kranidhi in 1973 (Petronoti 1985:79). While some of these sailors eventually settle permanently in the capital, many return to invest their earnings in a new house and economic operations in their home villages. Several who were interviewed said they came back because there was *dhouleia* (work, or means of economic survival) in the area, and therefore they did not have to move to Athens.

Commercial fishing, which began developing in the late nineteenth and early twentieth centuries, is the second seagoing enterprise currently of some importance to the area (Petronoti 1985:72) (Plate 12). Small boats fish for octopus, mullet, and a variety of small fish locally, and larger ones go after pelamyd tuna and other large fish in deeper waters as far away as Evvia and the Cyclades (Bintliff 1977). The Argosaronic area commands over 15% of the Greek fishing production (Kolodny 1974:1:302). Some of this catch was sent to Athens or dropped off there from the beginning; new refrigeration techniques have made this even easier. The rest of the catch is now sold to the many local restaurants serving the tourist trade.

Cash cropping also increased in the southern Argolid after the shipping crisis (Plate 11).[4] Some former shipowners invested in their agricultural estates (Petronoti 1985:72), but much

TABLE 4

COMPARISON OF POPULATION LEVELS IN THE SOUTHERN ARGOLID, IDHRA, AND SPETSAI

Date	*S. Argolid*	*Spetsai*	*Idhra*
1829	6,765[a]	9,000[b]	16,500[c]
1851	10,343[d]	8,929[d]	12,325[d]
1870	10,677	8,443	7,428
1907	11,803	4,373	5,695
1920	10,540	3,243	3,409
1940	11,722	3,633	3,780
1961	11,958	3,378	2,794
1981	12,222	3,729	2,732

[a]Houliarakis 1973:1:40
[b]Estimate based on Houliarakis 1973:1:40
[c]Michaelides 1967:26
[d]Houliarakis 1973:2:4
All other figures are from Greek national census reports.

---

of this expansion was undertaken by small land owners, especially after various changes in land tenure occurred. Many fields and grazing areas were created near Fourni after 1864 when public and monastic lands were given to those who fought in the revolution (Gavrielides 1976b:145). The Voulgaris estate near Ermioni was divided and sold following 1929, as were the Spetsiote estates near Kosta. The Idhriote *metochi* is now rented out for seasonal laborers, and the lands of the Avgo monastery were distributed to the northern villagers in the 1930s (Plate 8). Finally, several areas described as uncultivated in the nineteenth century (Curtius 1852:2:416-465; Philippson 1892:30-58) were opened up to agriculture in the twentieth century. The lands around Iliokastro and those of the Thermisi plain are prime among them.

The change to commerical agriculture is apparent not only from the increased extent of cultivated areas in the southern Argolid but also the nature of the crops that are grown. Although olive oil from Kranidhi is still considered of high quality in Athenian shops, it brings significant profits for only a few families (Petronoti 1985:79). Several other crops have become as important as olives to the area. Citrus orchards began to appear near Ermioni in the late nineteenth century (Paraskevopoulos 1895:49; Philippson 1892:48). Such orchards were extended further south to Kineta in the 1920s and begun in Fourni in the 1930s (Gavrielides 1976c:266) and the Thermisi plain after it was drained a little later. Increased irrigation has played a significant role in this (Harper 1976). Improved road communications and the expanded tourist market have even more recently spurred garden farming of fruits and vegetables. Small trucks from Radhon, for example, now regularly bring such produce down to the lower villages.

Many of the southern Argolid's shepherds have also turned to cash-producing activities (Plate 10). Cheese and other pastoral products are marketed as widely as Athens. Farming has grown as an activity in addition to or replacing herding. Dhidhima's shepherds now grow potatoes and cauliflower in the plain surrounding the village, while sheep and goats are kept in the nearby mountains (Koster and Koster 1976; Chang 1984). Dhidhima also spawned three new upland villages in the nineteenth and early twentieth century: Pelei, Loukaiti, and Iliokastro. Here they similarly engaged in a mixed shepherding-farming economy. Like the former shepherds of Radhon, they use the newly paved roads to bring their produce to market. The chalcopyrite mines near Iliokastro provided additional employment for this population for much of the twentieth century. Much of the restaurant and housekeeping staff at the large hotels is also drawn from these northern villages.

Local shopkeeping and crafts have been affected by the southern Argolid's increased involvement in external economic systems. Traditional crafts are on the decline. For example, looms and knowledge of manual weaving techniques are disappearing (J. Koster 1976:35), and while the only potter remaining in Ermioni changed production strategies to meet new demands, he still believes his craft will die with him (Kardulias, forthcoming). At the same time, local shops have increasingly marketed goods produced outside the southern Argolid. During the nineteenth century, such items as metals, wood, cloth, sugar, and wheat were imported from Athens and abroad, as *pantopoleia* (general stores) were stimulated by the prosperity of shipping (Petronoti 1985:69). Since then the amount and variety of externally produced items has greatly increased. Every settlement with over 30 residents has at least one *pantopoleion* stocked with such wares, and the larger villages have clusters of such shops.

The recent development of tourism in the coastal villages has also begun to provide an economic stimulus to the entire region similar to the shipping boom of the eighteenth and nineteenth centuries. Wealthier Athenians started building summer houses on Idhra and Spetsai in the 1880s, and the mainland has provisioned some of this seasonal trade since that time. Spurred by financial supports from the National Tourist Organization and the Greek Bank for Industrial Development (Kolodny 1974:1:423-432), large hotels also began to appear at Porto Kheli in the 1960s. Others have since been built near Kosta, Ermioni, and along the Thermisi coast. Most are not owned by local families, nor is their technical staff drawn from the region. They do however, employ local villagers for construction, housekeeping, and waiting on tables in their restaurants. Their existence has sparked other developments, such as smaller hotels and rooming houses, restaurants, tourist shops, and yacht facilities. Land in the southern coastal areas is also now sold for summer houses for middle- and upper-class Athenians. All this has brought an influx of outside capital and a need for local produce and services unmatched since the shipping boom. It is of little surprise that residents of Porto Kheli believe tourism is what stemmed the wave of migration away from the village that had begun in the 1950s.

When migrants from Porto Kheli returned in the 1960s to partake in the tourist boom, they were repeating another pattern that has kept the southern Argolid economically viable. Since the second half of the nineteenth century, many residents have left the area for a while and then returned to invest their earnings in farming or other local operations (Gavrielides 1976c). The merchant marine has long served this purpose. Some other temporary migration streams were directed toward construction of the Suez and Corinth Canals, and work in Syros in the nineteenth century (Petronoti 1985:76). Even migration to the United States and Canada was largely cyclical until World War II (Petronoti 1985:77). Such return-migration patterns signal the southern Argolid's ability to hold its population, a topic to which this chapter now turns.

MODERN SETTLEMENT PATTERNS

Such is the economic history that explains recent changes in settlement patterns in the southern Argolid. The transformations of the last one hundred fifty years have greatly influenced people's residential choices. Two major demographic trends have resulted. First,

TABLE 5

REGIONAL DEMOGRAPHIC HISTORY OF THE SOUTHERN ARGOLID

Date	*Eparchy*	*Demos Kranidhi*	*Demos Ermioni*	*Demos Dhidhima*
1705	1,900[a]	1,300[a]	400[a]	200[a]
1829	6,765[b]	4,985[b]	1,226[b]	554[b]
1870	10,677	7,413	2,011	1,036
1879	10,005	6,705	2,097	1,253
1889	10,151	6,442	2,396	1,313
1896	12,666	8,236	2,802	1,628
1907	11,803	7,588	2,620	1,595
1920	10,540	5,962	3,099	1,497
1928	9,893	5,813	2,361	1,719
1940	11,722	6,523	2,868	2,331
1951	11,903	6,267	3,322	2,314
1961	11,958	6,024	3,507	2,427
1971	11,794	6,055	3,341	2,398
1981	12,222	6,317	3,651	2,254

[a]Topping 1976
[b]Houliarakis 1973:1:40
All other figures are from the Greek national census reports. (The figures for the three original *demoi* (towns) were based on their original boundaries. All three *demoi* have since been designated as *koinotites* (communes), and new settlements have split off from them.)

the eparchy as a whole has increased its population levels since the Greek Revolution (Table 5). While housing less than 2,000 people in the early eighteenth century, the southern Argolid grew to over 6,700 by Greek independence, maintained a level of around 10,000 throughout the nineteenth century, and rose to over 12,000 in 1981. Kranidhi was the only settlement experiencing significant loss during this time, and even this did not affect the region's overall population levels. Second, there has been a proliferation of settlements in the southern Argolid in the modern period. While there were only four villages in the southern Argolid in the late eighteenth century, there are over twenty today. Both of these developments reflect the economic history presented above and are now discussed in more detail.

At the time of Greek independence, Kranidhi was the largest settlement and center of the region. A well protected hill town, it had been the agricultural and administrative hub during the Turkish period. Its early eighteenth century population of 1,000 represented 53% of the southern Argolid's total population. During the shipping boom, Kranidhi gained new importance, attracting migrants from outside the area and growing to 4,813 by independence and over 7,000 shortly thereafter (Table 6). At this point Kranidhi housed over 95% of the southern Argolid's population. Kranidhi's demographic history changed with the shipping crisis of the late nineteenth century, however (Plate 9a). Almost continual out-migration has brought it to its 1980 level of just under 4,000 people (or 58% of the eparchy). That all the censuses between 1879 and 1940 show between 1,000 and 2,000 Kranidhi citizens residing elsewhere reinforces the role of migration in the town's recent history. Although some migration went to other settlements within the eparchy, much was to Athens or abroad.

TABLE 6

POPULATION HISTORY OF SETTLEMENTS NEAR FRANCHTHI CAVE

Date	*Kiladha*	*Fourni*	*Kranidhi*
1705	no figure given	260[a]	1,000[a]
1829	16[b]	156[c]	4,813[c]
1879	231	319	5,628
1889	348	347	5,500
1896	464	269	6,954
1907	641	448	6,033
1920	566	470	4,384
1928	631	403	4,214
1940	815	463	4,588
1951	866	432	4,280
1961	884	422	3,942
1971	983	372	3,724
1981	1,062	371	3,794

[a]Gavrielides 1976c
[b]Topping 1976
[c]Houliarakis 1973:1:40
All other figures are from the Greek national census reports.

In sum, eighteenth-century Kranidhi underwent the same boom experienced by Idhra and Spetsai, growing disproportionately to the rest of the eparchy. As shipping declined and economic activities became more diversified within the region, however, it fell back to its former position. It is currently the capital of the eparchy and the major commercial center. Most of its over 125 shops market goods that come from outside the region, thus making it the major distribution point for larger economic systems. How long it remains the center depends on how coastal settlements such as Ermioni or Porto Kheli now grow.

While Kranidhi has experienced both ups and downs during the modern period, its two harbors have witnessed steady growth. During the shipping boom, both Porto Kheli and Kiladha were used to harbor Kranidhiote ships. Porto Kheli's well-protected harbor and proximity to Spetsai made it the primary port (Miliarakis 1886:227; Paraskevopoulos 1895:58).[5] This account focuses, however, on Kiladha, the settlement closest to Franchthi Cave. Kiladha began the modern period as a tiny hamlet clustered around its monastery of 16 monks (Houliarakis 1973:1:29). Standing in clear sight of Kranidhi, it served as a Kranidhiote harbor and seasonal fishing site. Kiladha steadily grew during the nineteenth century, however, as a center for some of the area's new economic activities (Table 6, Plate 9b). Surpassing 200 people by 1879, it has more than 1,000 today.

Shipbuilding, fishing, and the merchant marine have provided these new residents of Kiladha with their livelihood (Antonakatou and Mavros 1973:143; Petronoti 1985:66). Various census compilations show only about 20% of the village's population engaged in farming, something seconded in residents' opinions that they live in a seafaring village. The southern Argolid diversified its maritime activities after Greek independence. In the absence of warfare and piracy, residence on the coast thus provided more advantages than residence in a hill town. Kiladha's large harbor had long been valued, and its population thus grew. Local records reveal that two-thirds of this population moved to Kiladha from Kranidhi, while most of the remaining settlers came from outside the region, especially Leonidhion and Navplion. The Kranidhiote migrants generally held land or had other prior connections to Kiladha. Those from other regions were often either spouses encountered in long-distance sea expeditions or fishermen looking for a more economically viable base.

Agricultural settlements also proliferated in the southern Argolid during the modern period. At least three zones can be identified. First, the shepherding-farming village of Dhidhima grew dramatically, from under 200 in the early eighteenth century to over 1,600 in 1896. Groups of people from Dhidhima then created the three new satellite villages of Pelei, Loukaiti, and Iliokastro. Dhidhima's size correspondingly dropped to 1,217 people by 1981, but the combined population of all four villages in this zone rose to 2,254. Second, the transhumant Valtetsiote shepherds who had long used the coast east of Thermisi for winter pasturage began to settle there permanently in large numbers after World War II (Plate 10d). There they turned to commercial agriculture and tourism. This zone grew so rapidly that it became a separate *koinotis* (local administrative unit) in 1956. Third, the area east and south of Ermioni has witnessed an increase in isolated farmsteads and small hamlets (Murray and Kardulias 1986) connected with the spread of citrus orchards and other cash crops. This population was drawn both from Ermioni and more remote areas of the Peloponnese.

The two farming settlements closest to Franchthi Cave reflect such agricultural developments. Kambos is a non-nucleated scatter of houses situated midway along the main road between Kiladha and Fourni. This location gives its residents access to the fields of the Kiladha plain. Its current population of 98 was slowly built up over the last century as part of the area's general agricultural dispersion. Fourni, on the other hand, is one of the four original eighteenth century settlements of the region. Its 1829 population of 156 people steadily increased to 463 in 1940, but it has tapered off slightly since then (Table 6). Gavrielides (1976b) connects Fourni's growth to the successful development of a mixed economy of agriculture, seafaring, sponging, and migration. This kept its original population in place and attracted transhumant shepherds to join the village in the twentieth century (Gavrielides 1976c:265).

Overall, the southern Argolid has thus successfully negotiated with the new systems of modern Greece and thereby remained economically and demographically stable. The eparchy has almost always experienced positive rates of intercensal population growth since Greek independence. This contrasts with the many areas of rural Greece which have been depopulated (Baxevanis 1972; Kolodny 1974; Sutton 1983), and even with the nearby islands of Idhra and Spetsai (Table 4). The region's stability does not reflect spontaneous, indigenous development so much as the role it has played in larger systems. The southern Argolid has occupied an intermediary position in the highly centralized, externally dependent political economy of modern Greece. Subsistence activities have been replaced by those producing specialized goods and services for the centers of this system. The capital formation engendered by the shipping boom of the late eighteenth century and the connections the area had with the new national Greek government gave it an initial boost in this direction. A variety of lesser economic activities then kept it going. This is a precarious and lopsided balance, to be sure, something underscored by the many economic shifts the area has undergone in response to external changes. Throughout, however, the southern Argolid has maintained much of its population and has even attracted people from less advantageous areas of the Peloponnese. With little to fear from either pirates or imperial overlords and with greater access to land and capital than before, this population has dispersed widely throughout the region.

CONTRASTS AND CONTINUITIES WITH ANTIQUITY

While the analysis just presented places modern settlement patterns in the context of modern political economy, it also illuminates the southern Argolid's more distant past. Comparisons with previous periods of the area's history reveal several constants identifying the region's particular character. Jameson (1976) and van Andel (this volume) have recognized the great importance of the sea in framing the southern Argolid's lifeways. Van Andel et al. (1986) believe the basic categories of economic pursuit there have largely remained unchanged since the Neolithic: seafaring, fishing, pastoralism, and farming. The dryness of the environment has also been a constant negative factor in the region's agriculture (e.g., van Andel and Vitaliano, this volume).

Such comparisons between past and present, however, also reveal significant changes. Whereas the Franchthi site was the only Neolithic settlement in the southern Argolid until fairly late in the period (Jacobsen 1981:311), there are now over twenty villages in addition to scattered farmsteads and hamlets. This seems to be part of a cyclical pattern between times when there are a few small settlements in the southern Argolid and times when there are many settlements and larger populations. Because such a transition occurred in the modern period, the explanations given in this chapter may be useful in understanding what brings that cycle about.

The transition from low to high population levels seems to reflect changes in the southern Argolid's external connections. Bintliff's idea (1977) that there are only four good agricultural zones in the region (Dhidhima, Fourni, Kranidhi, and Ermioni) and that in times of isolated, locally self-sufficient systems population will be concentrated in them is an accurate description of early eighteenth century settlement patterns. Subsequent events have involved the southern Argolid in production for external systems, and a different settlement pattern has resulted. Jameson et al. (forthcoming) have combined this understanding of the present with knowledge of similar periods in antiquity to propose that settlements proliferate and more marginal lands are brought under cultivation in the southern Argolid only when such conditions prevail. They convincingly argue that the region cannot support a large population on subsistence agriculture, but only when considerable trade is occurring. Because of the region's seafaring traditions, the southern Argolid has often been important when Greek maritime commerce increases. At those times, it attracts population, both labor and capital imputs grow, and agricultural production increases. This has certainly been true for the modern period. The very significant capital accumulation at this time plus the new technology of mechanized agriculture have allowed the highest population levels that the area has probably ever experienced.

The area around Franchthi Cave stands at a point of juncture in this dynamic system. It is midway between mountains and lowlands in a location suitable for both agriculture and maritime activities. Thus a degree of constant habitation in this general vicinity is to be expected. It provides sufficient agricultural land for subsistence farming, and it abuts a good harbor for access to the outside. Shepherding, farming, gathering, fishing, and seafaring are all suitable to the area and provide a versatile base. The main agricultural settlement has moved back from the now sea-locked cave to the fertile Fourni valley. The main maritime settlement has moved across the harbor to the coastal position with greatest access to Kranidhi. Another agricultural settlement, Kambos, has recently appeared. While these developments contrast with certain periods of the area's past, they also can be seen as the latest phase in a cyclical pattern that has long characterized the Franchthi area.

NOTES

1. The ''southern Argolid'' here refers to the area encompassed by the boundaries of the present Eparchy of Ermionis (Hermionid) (Figure 28).

2. My work in the southern Argolid was undertaken as part of the Argolid Exploration Project led by M. Jameson, Tj. van Andel, and C. Runnels. H. Forbes and I were the project ethnologists, with Forbes concentrating on agricultural production and me on settlement patterns. I spent the summers of 1981 and 1982 in the region, being supported during the latter by a Summer Stipend from the National Endowment for the Humanities. During these field seasons I collected data from local records and interviewed individuals in each of the region's villages. Village officals were remarkably gracious and open in answering my questions and providing information. In analyzing the data collected, I was ably assisted by S. Langdon.

3. The Idhriotes struggled against consolidation of power (Petropulos 1968:119-123). An early rebellion on their part only resulted in their having to accept but one-third of the war reparations they had requested from the now angry central government.

4. H. Forbes is currently analyzing agricultural statistics for the southern Argolid to produce a more detailed understanding of its recent agricultural history.

5. Porto Kheli's population also grew greatly during the modern period, from almost nothing at the start of this time to 129 in 1879 and 754 in 1981.

REFERENCES CITED

Antonakatou, D., and T. Mavros

1973 *Argolidhos Periyisis.* Nomarchia Argolidhos, Navplion.

Aranitis, S. A.

1963 Die Entstehung der Eruptivgesteine vom Hermionigebiet und die mit Ihnen verbundene Vererzung (in Greek with German summary). *Annales Géologiques des Pays Helléniques* 14: 213- 304.

Aronis, G. A.

1938 Die Eruptivgesteine der Umgebung der Erzlagerstätte Karakassi (Ermionis) (in Greek with German summary). *Praktika Akademias Athenon* 13:481-487.

1951 Research on the Iron-pyrite Deposits in the Hermioni Mining District (in Greek with English summary). *Geological and Geophysical Studies,* Department of Subsurface Research, Ministry of Coordination, Athens, pp. 153-188.

Bachmann, Gerhard Heinz, and Hans Risch

1976 Ein oberjurassisch-unterkretazischer (eohellenischer) Flysch in der Argolis und der Bau der Lighourion-Mulde (Peloponnes, Griechenland). *Neues Jahrbuch für Geologie und Paläontologie, Abhandlungen* 152:137-160.

1978 Late Mesozoic and Paleogene Development of the Argolis Peninsula (Peloponnesos). In *Alps, Apennines, Hellenides,* edited by H. Closs, D. Roeder, and K. Schmidt, pp. 424-427. Inter-Union Commission on Geodynamics, Scientific Reports, vol. 38. Schweizerbart, Stuttgart.

1979 Die geologische Entwicklung der Argolis-Halbinsel (Peloponnes, Griechenland). *Geologisches Jahrbuch,* Reihe B 32:3-177.

Bannert, Dieter, and Hans Bender

1968 Zur Geologie der Argolis-Halbinsel (Peloponnese, Griechenland). *Geologica et Palaeontologica* 2:151-162.

Baxevanis, John J.

1972 *Economy and Population Movements in the Peloponnesos of Greece.* National Centre of Social Research, Athens.

Berger, A., J. Imbrie, J. Hays, G. Kukla, and B. Saltzman (editors)

1984 *Milankovitch and Climate: Understanding the Response to Astronomical Forcing.* NATO ASI Series. Series C. Mathematical and Physical Sciences 226. Reidel, Dordrecht.

Berger, W. H.

1977 Deep-Sea Carbonate and the Deglaciation Preservation Spike in Pteropods and Foraminifera. *Nature* 269:301-304.

Berger, W. H., and J. S. Killingley

1982 Box Cores from the Equatorial Pacific: 14-C Sedimentation Rates and Benthic Mixing. *Marine Geology* 45:93-125.

Berger, W. H., J. S. Killingley, and E. Vincent

1985 Timing of Deglaciation from an Oxygen Isotope Curve for Atlantic Deep-Sea Sediments. *Nature* 314:156-158.

Bintliff, John L.
1976a Sediments and Settlement in Southern Greece. In *Geoarchaeology: Earth Science and the Past,* edited by D. A. Davidson and M. L. Shackley, pp. 267-275. Duckworth, London.
1976b The Plain of Western Macedonia and the Neolithic Site of Nea Nikomedia. *Proceedings of the Prehistoric Society* 42:241-262.
1977 *Natural Environment and Human Settlement in Prehistoric Greece.* BAR Supplementary Series 28. British Archaeological Reports, Oxford.

Birkeland, Peter W.
1984 *Soils and Geomorphology.* Oxford University Press, New York.

Bloom, A. L.
1977 *Atlas of Sea Level Curves.* International Geological Correlation Programme, Project 61, Cornell University, Ithaca, New York.
1983 Sea Level and Coastal Geomorphology of the United States through the Late Wisconsin Glacial Maximum. In *The Late Pleistocene,* edited by Stephen C. Porter, pp. 215-229. Late-Quaternary Environments of the United States, vol. 1, H. E. Wright, Jr., general editor. University of Minnesota Press, Minneapolis, Minnesota.

Bloom A. L., W. S. Broecker, J. M. A. Chappell, R. K. Matthews, and K. J. Mesolella
1974 Quaternary Sea Level Fluctuations on a Tectonic Coast: New 230-Th/234-U Dates from the Huon Peninsula, New Guinea. *Quaternary Research* 4:185-205.

Bottema, S.
1979 Pollenanalytical Investigations in Thessaly, Greece. *Palaeohistoria* 21:19-40.

Bowen, D. Q.
1978 *Quaternary Geology: A Statistical Framework for Multidisciplinary Work.* Pergamon Press, Oxford.

Brakenridge, G. Robert
1980 Widespread Episodes of Stream Erosion during the Holocene and their Climatic Cause. *Nature* 283:655-656.

Butzer, K. W.
1964 *Environment and Archaeology: An Ecological Approach to Prehistory.* Aldine, Chicago.
1982 *Archaeology as Human Ecology: Method and Theory for a Contextual Approach.* Cambridge University Press, Cambridge.

Chang, C.
1984 The Ethnoarchaeology of Herding Sites in Greece. *MASCA Journal* 3:44- 48.

Chappell, J. M.
1974 Geology of Coral Terraces, Huon Peninsula, New Guinea: A Study of Quaternary Tectonic Movements and Sea Level Changes. *Geological Society of America, Bulletin.* 85:553-570.

Clark, J. A., W. E. Farrell, and W. R. Peltier
1978 Global Changes in Postglacial Sea Level: A Numerical Calculation. *Quaternary Research* 9:265-287.

CLIMAP Project Members
1976 The Surface of the Ice-Age Earth. *Science* 191:1131-1137.

Cline, R. M., and J. D. Hays (editors)
1976 *Investigation of Late Quaternary Paleoceanography and Paleoclimatology.* Geological Society of America, Memoir 145. Geological Society of America, Boulder, Colorado.

Cramp, A., M. B. Collins, S. J. Wakefield, and F. T. Banner
1984 Sapropelic Layers in the N.W. Aegean Sea. In *The Geological Evolution of the Eastern Mediterranean,* edited by J. E. Dickson and A. H. F. Robertson, pp. 807-814. Geological Society of London, Blackwell, London.

Cronin, Thomas M.
1983 Rapid Sea Level and Climate Change: Evidence from Continental and Island Margins. *Quaternary Science Reviews* 1:177-214.
Curray, Joseph R.
1960 Sediments and History of Holocene Transgression, Continental Shelf, Northwest Gulf of Mexico. In *Recent Sediments, Northwest Gulf of Mexico,* edited by F. P. Shepard, F. B. Phleger, and Tj. H. van Andel, pp. 221-266. American Association of Petroleum Geologists, Tulsa, Oklahoma.
1964 Transgressions and Regressions. In *Papers in Marine Geology in Honor of Francis P. Shepard,* edited by R. L. Miller, pp. 177-203. McMillan, New York.
Curray, Joseph R., and David G. Moore
1963 Facies Delineation by Acoustic-Reflection: Northern Gulf of Mexico. *Sedimentology* 2:130-14.
Curtius, Ernst
1852 *Peloponnesos: Eine historisch-geographische Beschreibung der Halbinsel,* vol. II. Perthes, Gotha.
Davidson, Donald A.
1980 Erosion in Greece during the First and Second Millennia B.C. In *Timescales in Geomorphology,* edited by R. A. Cullingford, D. A. Davidson, and J. Lewin, pp. 143-158. John Wiley, New York.
Davis, Richard A., Jr. (editor)
1978 *Coastal Sedimentary Environments.* Springer, New York.
Delmas, Robert J., Jean-Marc Ascencio, and Michel Legrand
1980 Polar Ice Evidence that Atmospheric CO_2 20,000 Years B.P. Was 50% of Present. *Nature* 284:155-157.
Dennell, Robin
1983 *European Economic Prehistory: A New Approach.* Academic Press, New York.
Denton, George H., and Wibjörn Karlén
1973 Holocene Climatic Variations: Their Pattern and Possible Cause. *Quaternary Research* 3:155-205.
Dietrich, G., K. Kalle, W. Kraus, and G. Siedler
1975 *Allgemeine Meereskunde.* 3rd edition. Bornträger, Berlin.
Dodge, Richard E., Richard G. Fairbanks, Larry K. Benninger, and Florentine Maurasse
1983 Pleistocene Sea Levels from Raised Coral Reefs of Haiti. *Science* 219:1423-1425.
Drost, Brian William
1974 *Late Quaternary Stratigraphy of the Southern Argolid (Peloponnese, Greece).* Unpublished M.Sc. thesis, Geology Department, University of Pennsylvania, Philadelphia.
Dürr, St., R. Altherr, J. Keller, M. Okrusch, and E. Seidel
1978 The Median Aegean Crystalline Belt: Stratigraphy, Structure, Metamorphism, Magmatism. In *Alps, Appenines, Hellenides,* edited by H. Closs, D. Roeder, and K. Schmidt, pp. 455-476. Inter-Union Commission on Geodynamics, Scientific Reports, vol. 38. Schweizerbart, Stuttgart.
Dufaure, Jean-Jacques, Bernard Bousquet, and Pierre-Yves Pechoux
1979 Contributions de la géomorphologie à la connaissance du Quaternaire continental Grec, en relation avec les études de néotectonique. *Revue de Géologie Dynamique et Géographie Physique* 21:29-40.
Duplessy, J. C., G. Delibrias, J. L. Turon, C. Pujol, and J. Duprat
1981 Deglacial Warming of the Northeastern Atlantic Ocean: Correlation with the Paleoclimatic Evolution of the European Continent. *Palaeogeography, Palaeoclimatology, Palaeoecology* 35: 121-144.
Fairbridge, R. W.
1961 Eustatic Changes in Sea Level. In *Physics and Chemistry of the Earth,* vol. 4, edited by L. C. Ahrens, pp. 99-185.

Finlay, George

1877 *A History of Greece From Its Conquest By the Romans to the Present Time, B.C. 146-A.D. 1864.* 6 vols. Clarendon Press, Oxford.

Flemming, N. C.

1968 Holocene Earth Movements and Eustatic Sea Level Change in the Peloponnese. *Nature* 217:1031-1032.

1978 Holocene Eustatic Changes and Coastal Tectonics in the Northeast Mediterranean: Implications for Models of Crustal Consumption. *Philosophical Transactions of the Royal Society of London* series A, 289:405-458.

Forbes, Hamish A.

1976 "We have a Little of Everything:" The Ecological Basis of Some Agricultural Practices in Methana, Trizinia. In *Regional Variation in Modern Greece and Cyprus: Toward a Perspective on the Ethnography of Greece,* edited by Muriel Dimen and Ernestine Friedl, pp. 236-250. Annals of the New York Academy of Sciences, vol. 268, New York.

1982 *Strategies and Soils: Technology, Production, and Environment in the Peninsula of Methana.* Unpublished Ph.D. dissertation, University of Pennsylvania, Philadelphia.

Forbes, Hamish A., and Harold A. Koster

1976 Fire, Axe, and Plow: Human Influence on Local Plant Communities in the Southern Argolid. In *Regional Variation in Modern Greece and Cyprus: Toward a Perspective on the Ethnography of Greece,* edited by Muriel Dimen and Ernestine Friedl, pp. 109-126. Annals of the New York Academy of Sciences, vol. 268, New York.

Forbes, Mary Clark

1976a Farming and Foraging in Prehistoric Greece: A Cultural Ecological Perspective. In *Regional Variation in Modern Greece and Cyprus: Toward a Perspective on the Ethnography of Greece,* edited by Muriel Dimen and Ernestine Friedl, pp. 127-142. Annals of the New York Academy of Sciences, vol. 268, New York.

1976b The Pursuit of Wild Edibles, Present and Past. *Expedition,* 19:12-18.

Forney, G. G.

1971 *Geology of the Kranidhi Region, Argolid Peninsula, Greece.* Unpublished senior thesis, Geology Department, University of Pennsylvania, Philadelphia.

Gavrielides, Nicholas E.

1976a *A Study in the Cultural Ecology of an Olive-Growing Community: The Southern Argolid, Greece.* Unpublished Ph.D. dissertation, Indiana University, Bloomington.

1976b The Impact of Olive Growing on the Landscape of the Fourni Valley. In *Regional Variation in Modern Greece and Cyprus: Toward a Perspective on the Ethnography of Greece,* edited by Muriel Dimen and Ernestine Friedl, pp. 143-157. Annals of the New York Academy of Sciences, vol. 268, New York.

1976c The Cultural Ecology of Olive Growing in the Fourni Valley. In *Regional Variation in Modern Greece and Cyprus: Toward a Perspective on the Ethnography of Greece,* edited by Muriel Dimen and Ernestine Friedl, pp. 265-274. Annals of the New York Academy of Sciences, vol. 268, New York.

Gell, William, Sir

1810 *The Itinerary of Greece.* T. Payne, London.

Georgas, G. E.

1937 *Meleti Peri Spongon, Spongalieias kai Spongemboriou.* Akademias Athinon, Athens.

Gifford, John A.

1983 Core Sampling of a Holocene Marine Sedimentary Sequence and Underlying Neolithic Cultural Material off Franchthi Cave, Greece. In *Quaternary Coastlines and Marine Archaeology,* edited by P. M. Masters and N. C. Flemming, pp. 269-281. Academic Press, New York.

Grosswald, M. G.

1980 Late Weichselian Ice Sheet of Northern Eurasia. *Quaternary Research* 13:1-32.

Hansen, Julie M.

1980 *The Palaeoethnobotany of Franchthi Cave, Greece.* Unpublished Ph.D. dissertation, University of Minnesota, Minneapolis.

forthcoming *The Palaeoethnobotany of Franchthi Cave.* In Excavations at Franchthi Cave, Greece, T. W. Jacobsen, general editor. Indiana University Press, Bloomington and Indianapolis.

Harden, Jennifer W.

1982 A Quantitative Index of Soil Development from Field Descriptions: Examples from a Chronosequence in Central California. *Geoderma* 28:1-28.

Harper, D. B.

1976 "Just Add Water. . . ." *Expedition* 19:40-49.

Hassan, Fekri A.

1981 Historical Nile Floods and their Implications for Climatic Change. *Science* 212:1142-1145.

Haynes, C. V., Jr.

1968 Geochronology of the Late-Quaternary Alluvium. In *Means of Correlation of Quaternary Sequences,* edited by Herbert E. Wright, Jr., and Roger B. Morrison, pp. 591-631. University of Utah Press, Salt Lake City.

Herz, Norman, and Charles J. Vitaliano

1983 Archaeological Geology in the Eastern Mediterranean: A Symposium Report. *Geology* 11:49-53.

Herzfeld, Michael

1982 *Ours Once More: Folklore, Ideology, and the Making of Modern Greece.* University of Texas Press, Austin.

Houliarakis, Mihail

1973 *Geografiki, Dioikitiki kai Plithismiaki Exelixis tis Elladhos, 1821-1971.* 2 volumes. National Centre of Social Research, Athens.

IAPSO (Advisory Committee on Tides and Mean Sea Level)

1985 Changes in Relative Mean Sea Level. *EOS, Transactions, American Geophysical Union,* 66:754-756.

Imbrie, John, and Nilva G. Kipp

1971 A New Micropaleontological Method for Quantitative Paleoclimatology: Application to a Late Pleistocene Caribbean Core. In *The Late Cenozoic Glacial Ages,* edited by Karl K. Turekian, pp. 71-182. Yale University Press, New Haven, Connecticut.

Jacobsen, Thomas W.

1976 17,000 Years of Greek Prehistory. *Scientific American* 234(6):76-87.

1981 Franchthi Cave and the Beginning of Settled Village Life in Greece. *Hesperia* 50(4):303-319.

Jacobsen, Thomas W., and David M. Van Horn

1974 The Franchthi Cave Flint Survey: Some Preliminary Results (1974). *Journal of Field Archeology* 1:305-308.

Jacobsen, Thomas W., and William R. Farrand

1987 *Franchthi Cave and Paralia: Maps, Plans and Sections.* Excavations at Franchthi Cave, Greece, fasc. 1, T. W. Jacobsen, general editor. Indiana University Press, Bloomington and Indianapolis.

Jacobshagen, V.

1972 Die Trias der mittleren Ost-Ägäis und ihre paläogeographischen Beziehungen innerhalb der Helleniden. *Deutsche Geologische Gesellschaft, Zeitschrift* 123:445-454.

Jacobshagen, V., St. Dürr, F. Kockel, K.-Q. Kopp, and G. Kowalczyk

1978 Structure and Geodynamic Evolution of the Aegean Region. In *Alps, Apennines, Hellenides,* edited by H. Closs, D. Roeder, and K. Schmidt, pp. 537-564. Inter-Union Commission on Geodynamics, Scientific Reports, vol. 38. Schweizerbart, Stuttgart.

Jameson, Michael H.

1976 The Southern Argolid: The Setting for Historical and Cultural Studies. In *Regional Variation in Modern Greece and Cyprus: Toward a Perspective on the Ethnography of Greece,* edited by Muriel Dimen and Ernestine Friedl, pp. 74-91. Annals of the New York Academy of Sciences, vol. 268, New York.

Jameson, Michael H., Curtis N. Runnels, and Tjeerd H. van Andel

forthcoming *The Southern Argolid, a Greek Countryside from Prehistory to the Present Day.* Stanford University Press, Stanford.

Johnson, R. F.

1980 *One-centimeter Stratigraphy in Foraminiferal Ooze: Theory and Practice.* Unpublished Ph.D. dissertation, Scripps Institution of Oceanography, University of California, San Diego.

Karanikolas, P. K.

1980 *To Kranidhi.* Ieras Mitropoleos, Corinth.

Kardulias, P. Nick

forthcoming Pottery Production Near Modern Ermioni. In *Shepherds, Farmers, and Sailors: The Regional Ethnohistory of the Southern Argolid Peninsula,* edited by Susan Buck Sutton. Stanford University Press, Stanford.

Kelletat, D, G. Kowalczyk, and B. Schroeder

1976 A Synoptic View on the Neotectonic Development of the Peloponnesian Coastal Regions. *Deutsche Geologische Gesellschaft, Zeitschrift* 127:447-465.

Kennett, James P., and N. J. Shackleton

1975 Laurentide Ice Sheet Meltwater Recorded in Gulf of Mexico Deep-Sea Cores. *Science* 188:147-150.

Kidd, Robert B., Maria Bianca Cita, and William B. F. Ryan

1978 Stratigraphy of Eastern Mediterranean Sapropel Sequences Recovered During DSDP Leg 42A and their Paleoenvironmental Significance. *Initial Reports of the Deep Sea Drilling Project* 42(1):421-443.

Kidson, C.

1982 Sea Level Changes in the Holocene. *Quaternary Science Reviews* 1:121-151.

Knox, J. C.

1983 Responses of River Systems to Holocene Climates. In *The Holocene,* edited by H. E. Wright, Jr., pp. 26-41. Late-Quaternary Environments of the United States, vol. 2, H. E. Wright, Jr., general editor. University of Minnesota Press, Minneapolis, Minnesota.

Kolodny, Emile Y.

1974 *La population des îles de la Grèce. Essay de géographie insulaire en Mediterranée,* vols. 1-2. Edisud, Aix-en-Provence.

Kominz, M. A., G. R. Heath, T.-L. Ku, and N. G. Pisias

1979 Brunhes Time Scales and the Interpretation of Climatic Change. *Earth and Planetary Sciences Letters* 45:394-410.

Koster, Harold A., and Joan Bouza Koster

1976 Competition or Symbiosis? Pastoral Adaptive Strategies in the Southern Argolid, Greece. In *Regional Variation in Modern Greece and Cyprus: Toward a Perspective on the Ethnography of Greece,* edited by Muriel Dimen and Ernestine Friedl, pp. 275-285. Annals of the New York Academy of Sciences, vol. 268, New York.

Koster, Joan Bouza

1976 From Spindle to Loom: Weaving in the Southern Argolid. *Expedition* 19:29-39.

Kowalczyk, G.

1977 Jungquartäre Strandterrassen in SE-Lakonien, Peloponnes. *Proceedings of the VIth Colloquium on the Geology of the Aegean Region,* vol. 1, edited by G. Kallergis, pp. 435-445. Institute of Geological and Mining Research, Athens.

Kraft, John C.

1985 Marine Environments: Paleogeographic Reconstructions in the Littoral Region. In *Archaeological Sediments in Context,* edited by J. K. Stein and W. R. Farrand, pp. 111-125. Peopling of the Americas, vol. 1, Alan L. Bryan and Ruth Gruhn, general editors. University of Maine Press, Orono, Maine.

Kraft, John C., Stanley E. Aschenbrenner, and George Rapp, Jr.

1977 Paleogeographic Reconstructions of Coastal Aegean Archaeological Sites. *Science* 195:941-947.

Kraft, John C., Ilhan Kayan, and Oguz Erol

1980 Geomorphic Reconstructions in the Environs of Ancient Troy. *Science* 209:776-782.

Kraft, John C., George Rapp, Jr., and Stanley E. Aschenbrenner

1975 Late Holocene Paleogeography of the Coastal Plain of the Gulf of Messenia, and its Relationships to Archaeological Settings and to Coastal Change. *Geological Society America Bulletin* 86:1191-1208.

Kriezis, G. D.

1860 *Istoria tis Nisou Idhras.* A. S. Agapitou, Patras.

Ku, T.-L., and Z.-C. Liang

1983 The Dating of Impure Carbonates with Decay Series Isotopes. *Proceedings, Seminar on Alpha Particle Spectrometry and Low Level Measurement,* unnumbered. Harwell, England.

Kukla, George, and Madeleine Briskin

1983 The Age of the 4/5 Isotopic Stage Boundary on Land and in the Oceans. *Palaeogeography, Palaeoclimatology, Palaeoecology* 42:35-45.

Kutzbach, J. E.

1983 Monsoon Rains of the Late Pleistocene and Early Holocene: Patterns, Intensity, and Possible Causes of Changes. In *Variations in the Global Water Budget,* edited by F. A. Street-Perrott, M. Beran and R. Ratcliffe, pp. 361-370. Reidel, Boston.

Marius, C.

1976 *Study of the Vegetation of the Southern Part of the Argolis Peninsula.* Unpublished report of the Argolid Exploration Project.

Matthews, R. K.

1973 Relative Elevation of Late Pleistocene High Sea Level Stands: Barbados Uplift Rates and their Implications. *Quaternary Research* 3:147-153.

Meiggs, Russell

1982 *Trees and Timber in the Ancient Mediterranean World.* Clarendon Press, Oxford.

Miall, Andrew D.

1977 A Review of the Braided-River Depositional Environment. *Earth Science Reviews* 13:1-62.

1978 Lithofacies Types and Vertical Profile Models in Braided River Deposits: A Summary. In *Fluvial Sedimentology,* edited by Andrew D. Miall, pp. 597-604. Canadian Society of Petroleum Geologists, Memoir 5. Calgary, Canada.

Michaelides, Constantine E.

1967 *Hydra: A Greek Island Town, Its Growth and Form.* University of Chicago Press, Chicago.

Miliarakis, A.

1886 *Geografia politiki nea kai archaia tou nomou Argolidhos kai Korinthias.* Estias, Athens.

Miller, A. R., P. Tchernia, H. Charnock, and D. A McGill

1970 *Mediterranean Sea Atlas.* Woods Hole Oceanographic Institution Atlas Series 3. Woods Hole Oceanographic Institution, Woods Hole, Mass.

Mix, Alan C., and William F. Ruddiman

1984 Oxygen-Isotope Analyses and Pleistocene Ice Volumes. *Quaternary Research* 21:1-20.

Morley, Joseph J., and James D. Hays

1981 Towards a High-Resolution, Global, Deep-Sea Chronology for the Last 750,000 Years. *Earth and Planetary Science Letters* 53:279-295.

Morrison, R. B.
1976 Quaternary Soil Stratigraphic Concepts, Methods, and Problems. In *Quaternary Soils,* edited by W. C. Mahaney, pp. 77-108. Geo-Abstracts, Norwich, England.

Moussoulos, L.
1958 Les gisements pyriteux du district minier d'Hermione. *Annales Géologiques des Pays Helléniques* 9:119-155.

Mouzelis, Nicos P.
1978 *Modern Greece: Facets of Underdevelopment.* Holmes and Meier, New York.

Murray, Priscilla, and P. Nick Kardulias
1986 A Modern-Site Survey in the Argolid, Greece. *Journal of Field Archaeology* 13:21-41.

NID (Naval Intelligence Division)
1944 *Greece, Physical Geography, History, Administration, and Peoples.* Geographical Handbook Series BR 516, London.
1945 *Greece.* Geographical Handbook Series BR 516, London.

Nilsson, Tage
1983 *The Pleistocene: Geology and Life in the Quaternary Ice Age.* Reidel, Dordrecht.

Nockolds, S. R., W. B. O'B. Knox, and G. A. Chinner
1978 *Petrology for Students.* Cambridge University Press, Cambridge.

Orlandos, A. K.
1877 *Peri tis nisou Petsas i Spetson.* Rousopoulos, Piraeus.

Paraskevopoulos, D.
1895 *Taxeidia ana tin Elladha.* Korinnis, Athens.

Pastouret, L., H. Chamley, G. Delibrias, J.-C. Duplessy, and J. Thiede
1978 Late Quaternary Climate Changes in Western Tropical Africa Deduced from Deep-Sea Sedimentation off the Niger Delta. *Oceanologica Acta* 1:217-227.

Patton, Peter C., and Stanley A. Schumm
1981 Ephemeral-Stream Processes: Implications for Studies of Quaternary Valley Fills. *Quaternary Research* 15:24-43.

Payne, Sebastian
1975 Faunal Change at Franchthi Cave from 20,000 B.C. to 3,000 B.C. In *Archaeozoological Studies,* edited by A. T. Clason, pp. 120-131. Elsevier, Amsterdam.
1982 Faunal Evidence for Environmental/Climatic Change at Franchthi Cave (Southern Argolid, Greece), 25,000 B.P.-5000 B.P.: Preliminary Results (abstract). In *Palaeoclimates, Palaeoenvironments, and Human Communities in the Eastern Mediterranean Region in Later Prehistory,* edited by J. L. Bintliff and W. van Zeist. British Archaeological Reports, International Series 133:133-137.

Peltier, W. R.
1980 Models of Glacial Isostasy and the Relative Sea Level. In *Dynamics of Plate Interiors,* edited by A. W. Bally, et al., pp. 111-128. Geodynamic Series, vol. 1. American Geophysical Union, Washington, D.C.

Perlès, Catherine
1987 *Les Industries lithiques taillées de Franchthi.* Tome I, *Présentation générale et industries paléolithiques.* Excavations at Franchthi Cave, Greece, fasc. 3, T. W. Jacobsen, general editor. Indiana University Press, Bloomington and Indianapolis.

Perry, Allen
1981 Mediterranean Climate: A Synoptic Reappraisal. *Progress in Physical Geography* 5:107-113.

Peterson, G. M., T. Webb III, J. E. Kutzbach, T. van der Hammen, T. A. Wijmstra, and F. A. Street
1979 The Continental Record of Environmental Conditions at 18,000 yr B.P.: An Initial Evaluation. *Quaternary Research* 12:47-82.

Pethick, J.
1984 *An Introduction to Coastal Geomorphology.* Edward Arnold, London.

Petronoti, M.

1985 Schediasma yia ti meleti ton oikonomikon kai koinonikon schimatismon sto Kranidhi (1821-1981). *Greek Review of Social Research* 57:63-82.

Petropulos, John Anthony

1968 *Politics and Statecraft in the Kingdom of Greece 1833-1843.* Princeton University Press, Princeton.

Philippson, A.

1892 *Der Peloponnes.* Die griechischen Landschaften: Eine Landeskunde, Band III. R. Friedländer, Berlin.

1948 *Das Klima Griechenlands.* F. Dümmler Verlag, Bonn.

Picard, M. Dane, and Lee R. High, Jr.

1973 *Sedimentary Structures of Ephemeral Streams.* Developments in Sedimentology No. 17. Elsevier, Amsterdam.

Ponti, Daniel J., Dennis B. Burke, Denis E. Marchand, Brian F. Atwater, and Edward J. Helley

1980 Evidence for Correlation and Climatic Control of Sequences of Late Quaternary Alluvium in California. *Geological Society America, Abstracts with Programs* 12:501.

Pope, Kevin O., Curtis N. Runnels, and Teh-Lung Ku

1984 Dating Middle Palaeolithic Red Beds in Southern Greece. *Nature* 312:264-266.

Pope, Kevin O., and Tjeerd H. van Andel

1984 Late Quaternary Alluviation and Soil Formation in the Southern Argolid: Its History, Causes, and Archaeological Implications. *Journal of Archaeological Science* 11:281-306.

Pouqueville, F.-C.-H.-L.

1826 *Voyage de la Grèce.* 6 vols. Firman Didot, Paris.

Rackham, Oliver

1983 Observations on the Historical Ecology of Boeotia. *Annual of the British School at Athens* 78:291-351.

Rand McNally

1977 *The Rand McNally Atlas of the Oceans.* Rand McNally and Company, New York.

Raphael, C. Nicholas

1973 Late Quaternary Changes in Coastal Elis, Greece. *Geographical Review* 63:73-89.

Rossignol-Strick, Martine, Wladimir Nesteroff, Philippe Olive, and Colette Vergnaud-Grazzini

1982 After the Deluge: Mediterranean Stagnation and Sapropel Formation. *Nature* 295:105-110.

Ruddiman, William F., and J.-C. Duplessy

1985 Conference on the Last Deglaciation: Timing and Mechanism. *Quaternary Research* 23:1-17.

Ruddiman, William F., and Andrew McIntyre

1981a The North Atlantic Ocean During the Last Deglaciation. *Palaeogeography, Palaeoclimatology, Palaeoecology* 35:145-214.

1981b The Mode and Mechanism of the Last Deglaciation: Oceanic Evidence. *Quaternary Research* 16:125-134.

Runnels, Curtis N.

1981 *A Diachronic Study and Economic Analysis of Millstones from the Argolid, Greece.* Unpublished Ph.D. dissertation, Indiana University, Bloomington, Indiana.

1985 Trade and the Demand for Millstones in Southern Greece in the Neolithic and the Early Bronze Age. In *Prehistoric Production and Exchange: The Aegean and Eastern Mediterranean,* edited by A. Bernard Knapp and Tamara Stech, pp. 30-43. Monograph XXV. Institute of Archaeology, University of California, Los Angeles.

Sancetta, Constance, John Imbrie, and N. G. Kipp

1973 Climatic Record of the Past 130,000 Years in North Atlantic Deep-Sea Core V23-82: Correlation with Terrestrial Record. *Quaternary Research* 3:110-116.

Schumm, S. A.
1977 *The Fluvial System.* John Wiley, New York.
Shackleton, Judith C., and Tjeerd H. van Andel
1986 Prehistoric Environments, Shellfish Availability, and Shellfish Gathering at Franchthi, Southern Argolid, Greece. *Geoarchaeology* 1:127-143.
Shackleton, N. J.
1977 The Oxygen Isotope Stratigraphic Record of the Late Pleistocene. *Philosophical Transactions of the Royal Society of London,* series B. 280:169-182.
Shay, Jennifer M., and C. Thomas Shay
1978 Modern Vegetation and Fossil Plant Remains. In *Excavations at Nichoria,* vol. I, edited by George Rapp, Jr., and S. E. Aschenbrenner, pp. 41-59. University of Minnesota Press, Minneapolis.
Sheehan, Mark C.
1979 *The Postglacial Vegetational History of the Southern Argolid Peninsula, Greece.* Unpublished Ph.D. dissertation, Indiana University, Bloomington, Indiana.
Sheehan, Mark C., and K. B. Sheehan
1982 Floristic Survey and Vegetational Analysis of the Argolid Peninsula, Southern Greece. Ms. on file, Program in Classical Archaeology, Indiana University, Bloomington.
Spiridonakis, B. G.
1977 *Essays on the Historical Geography of the Greek World in the Balkans during the Turkokratia.* Institute for Balkan Studies, Thessaloniki.
Stanley, Daniel Jean
1978 Ionian Sea Sapropel Distribution and Late Quaternary Palaeoceanography in the Eastern Mediterranean. *Nature* 274:149-152.
Stanley, Daniel Jean, and Christian Blanpied
1980 Late Quaternary Water Exchange between the Eastern Mediterranean and the Black Sea. *Nature* 285:537-541.
Street-Perrott, F. A., and F. N. Roberts
1983 Fluctuations in Closed-Basin Lakes as an Indicator of Past Atmospheric Circulation Patterns. In *Variations in the Global Water Budget,* edited by F. A. Street-Perrott, M. Beran, and R. Ratcliffe, pp. 331-345. Reidel, Boston.
Strong, Frederick
1842 *Greece as a Kingdom.* Longman, Brown, Green, and Longmans, London.
Süsskoch, Hermann
1967 *Die Geologie der südöstlichen Argolis (Peloponnes, Griechenland).* Dissertation, Marburg.
Sutton, Susan Buck
1983 Rural-Urban Migration in Greece. In *Urban Life in Mediterranean Europe: Anthropological Perspectives,* edited by Michael Kenny and David I. Kertzer, pp. 225-249. University of Illinois Press, Urbana.
Thiede, Jörn
1978 A Glacial Mediterranean. *Nature* 276:680-683.
1980 The Late Quaternary Marine Paleoenvironment between Europe and Africa. In *Palaeoecology of Africa and the Surrounding Islands,* vol. 12, edited by E. M. van Zinderen Bakker and J. A. Coetzee, pp. 213-225. Balkema, Rotterdam.
Thunell, Robert C.
1979 Eastern Mediterranean Sea during the Last Glacial Maximum; An 18,000-Years B.P. Reconstruction. *Quaternary Research* 11:353-372.
Thunell, Robert C., and Douglas F. Williams
1982 Paleoceanographic Events Associated with Termination II in the Eastern Mediterranean. *Oceanologica Acta* 5:229-233.

Thunell, Robert C., and Douglas F. Williams

1983 Paleotemperature and Paleosalinity History of the Eastern Mediterranean During the Late Quaternary. *Palaeogeography, Palaeoclimatology, Palaeoecology* 44:23-39.

Thunell, Robert C., Douglas F. Williams, and James P. Kennett

1977 Late Quaternary Paleoclimatology, Stratigraphy, and Sapropel History in Eastern Mediterranean Deep-Sea Sediments. *Marine Micropaleontology* 2:371-388.

Tollner, H.

1976 Zum Klima von Griechenland. *Beiträge zur Landeskunde Griechenlands* 6:267-281.

Topping, Peter

1976 Premodern Peloponnesus: The Land and the People under Venetian Rule (1685-1715). In *Regional Variation in Modern Greece and Cyprus: Toward a Perspective on the Ethnography of Greece,* edited by Muriel Dimen and Ernestine Friedl, pp. 92-108. Annals of the New York Academy of Sciences, vol. 268, New York.

Turner, Judith, and James R. A. Greig

1975 Some Holocene Pollen Diagrams from Greece. *Review of Palaeobotany and Palynology* 20:171-204.

Vacalopoulos, Apostolos E.

1976 *The Greek Nation, 1453-1669.* Rutgers University Press, New Brunswick, New Jersey.

van Andel, Tjeerd H., Thomas W. Jacobsen, Jon B. Jolly, and Nikolaos Lianos

1980 Late Quaternary History of the Coastal Zone near Franchthi Cave, Southern Argolid, Greece. *Journal of Field Archaeology* 7(4):389-402.

van Andel, Tjeerd H., and Nikolaos Lianos

1983 Prehistoric and Historic Shorelines of the Southern Argolid Peninsula: A Subbottom Profiler Study. *International Journal of Nautical Archaeology and Underwater Exploration* 12: 303-324.

1984 High-Resolution Seismic Reflection Profiles for the Reconstruction of Post-Glacial Transgressive Shorelines: An Example from Greece. *Quaternary Research* 22:31-45.

van Andel, Tjeerd H., and Curtis N. Runnels

1987 *Beyond the Acropolis. A Rural Greek Past.* Stanford University Press, Stanford.

van Andel, Tjeerd H., Curtis N. Runnels, and Kevin O. Pope

1986 Five Thousand Years of Land Use and Abuse in the Southern Argolid, Greece. *Hesperia* 55:103-128.

van Andel, Tjeerd H., and Peter L. Sachs

1964 Sedimentation in the Gulf of Paria During the Holocene Transgression; a Subsurface Acoustic Reflection Study. *Journal of Marine Research* 22:30-50.

van Andel, Tjeerd H., and Judith C. Shackleton

1982 Late Paleolithic and Mesolithic Coastlines of Greece and the Aegean. *Journal of Field Archaeology* 9:445-454.

Van Horn, D. M.

1973 The Franchthi Cave Flint Survey. *A.F.F.A. News* 2(1):11- 13.

van Zeist, W., and S. Bottema

1982 Vegetational History of the Eastern Mediterranean and the Near East During the Last 20,000 Years. In *Palaeoclimates, Palaeoenvironments, and Human Communities in the Eastern Mediterranean Region in Later Prehistory,* edited by J. L. Bintliff and W. van Zeist. British Archaeological Reports, International Series 133:277-321.

Verheye, W., and M. T. Lootens-de Muynck

1974 A Study of the Present Landscape of the Fournoi Valley, Argolid, Greece, with Some Implications for Further Archaeological Research. *Société Belge d'Etudes Géographiques, Bulletin* 43:37-60.

Vita-Finzi, Claudio
1969 *The Mediterranean Valleys: Geological Changes in Historical Times.* Cambridge University Press, Cambridge.
Voreadis, G. D.
1958 Sur la genèse des gisements de pyrite et de manganite de l'Hermionide (Argolis) et sur leur rélations mutuelles (in Greek with French summary). *Hellenikes Geologikes Hetairias Deltion* 3:50-63.
Wagstaff, J. M.
1981 Buried Assumptions: Some Problems in the Interpretation of the "Younger Fill" Raised by Recent Data from Greece. *Journal of Archaeological Science* 8:247-264.
Washburn, A. L.
1980 *Geocryology.* John Wiley, New York.
Wendland, W. M.
1982 Geomorphic Responses to Climatic Forcing during the Holocene. In *Fluvial Geomorphology,* edited by M. Morisawa, pp. 299-310. State University of New York, Binghamtom, New York.
Wigley, T. M. L., and G. Farmer
1982 Climate of the Eastern Mediterranean and Near East. In *Palaeoclimates, Palaeoenvironments, and Human Communities in the Eastern Mediterranean in Later Prehistory,* edited by J. L. Bintliff and W. van Zeist. British Archaeological Reports, International Series 133:3-37.
Wijmstra, T. A.
1969 Palynology of the First 30 Metres of a 120 m Deep Section in Northern Greece. *Acta Botanica Neerlandica* 18:511-527.
Williams, Douglas F., and Thunell, Robert C.
1979 Faunal and Oxygen Isotopic Evidence for Surface Water Salinity Changes during Sapropel Formation in the Eastern Mediterranean. *Sedimentary Geology* 23:81-93.
Woillard, Geneviève M., and Willem G. Mook
1982 Carbon-14 Dates at Grande Pile: Correlation of Land and Sea Chronologies. *Science* 215: 159-161.
Wright, H. E., Jr.
1972 Vegetational History. In *The Minnesota Messenia Expedition: Reconstructing a Bronze Age Environment,* edited by William A. McDonald and George R. Rapp, Jr., pp. 188-199. University of Minnesota Press, Minneapolis.
Wright, L. D., and B. G. Thom
1977 Coastal Depositional Landforms. *Progress in Physical Geography* 1:412-459.

Plates

Plate 1. Woods of Aleppo pine on north-facing slope bordering a cultivated field in the Ermioni valley (*a*), and Kermes oak tree which has attained natural height and form in a dense hedgerow east of Kiladha (*b*). Note 1 m scale.

Plate 2. Juniper maquis typical of south- and west-facing slopes near Franchthi headland (*a*), and sparse, over-grazed and salt-pruned maquis near Kiladha (*b*); the larger shrubs are juniper and Kermes oak.

Plate 3. Phrygana with wild thyme and grasses in the Ermioni valley; note 1 m scale in foreground (*a*), and reeds and rushes on a coastal salt flat near Thermisi (*b*).

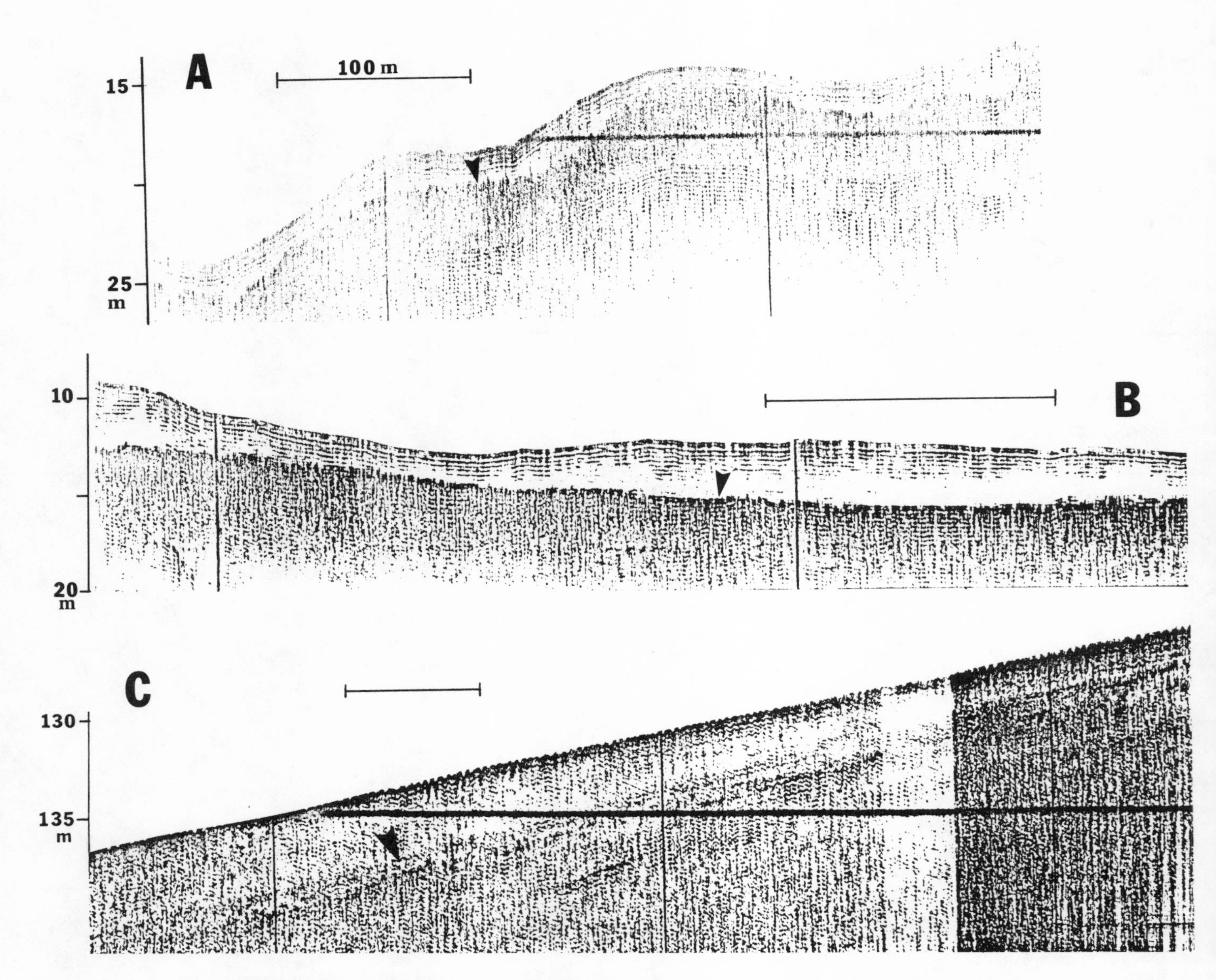

Plate 4. Seismic profiles illustrating bedrock reflector (*arrow in A*), basal alluvial plain reflector (*arrow in B*), and the marine slope deposits fringing the deepest glacial shore (*C*). Horizontal bars are 100 m scales. Vertical scales in meters below present sea level. Vertical exaggerations 9.5× (*A*), 19× (*B*), and 6.5× (*C*).

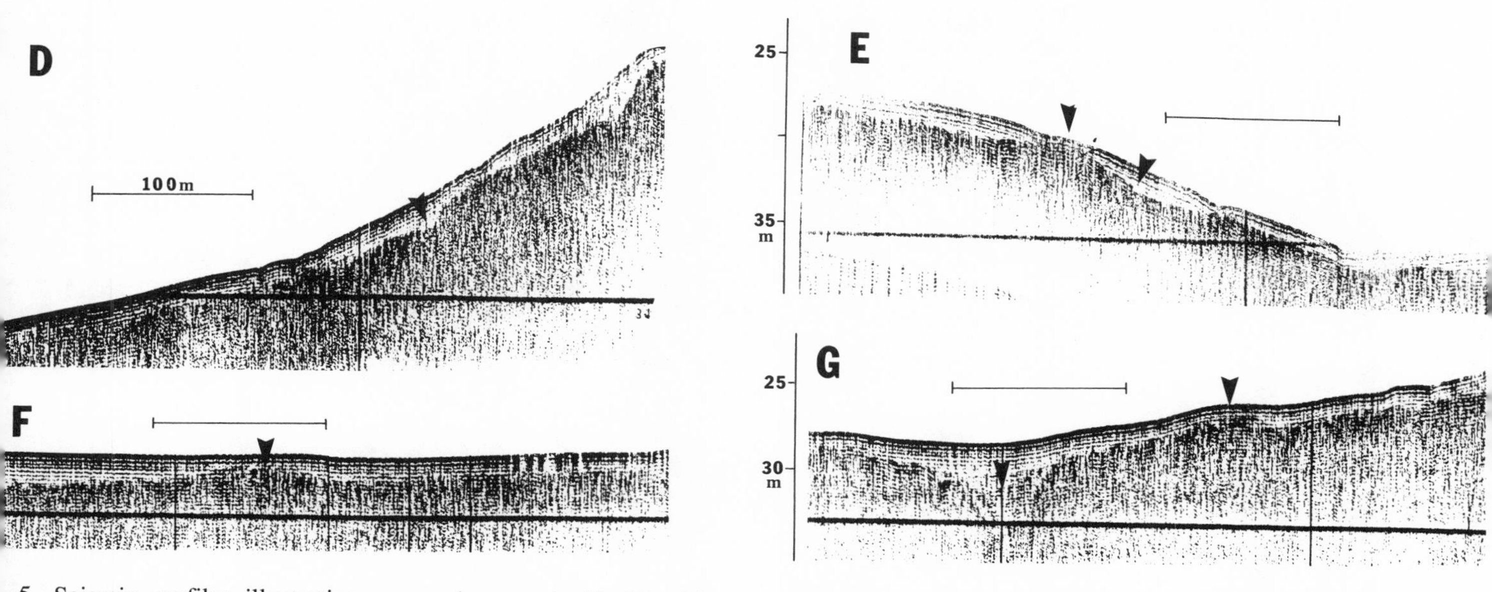

5. Seismic profiles illustrating scarps (*arrows in D, E*), ridges (*in F, in G on right*), and channels (*in G on left*). Note ridge left of scarp in *E*. Horizontal bars are 100 m scales. Vertical scales in meters below present sea level. Vertical exaggerations *D*), 9.5× (*E, F*), and 10× (*G*).

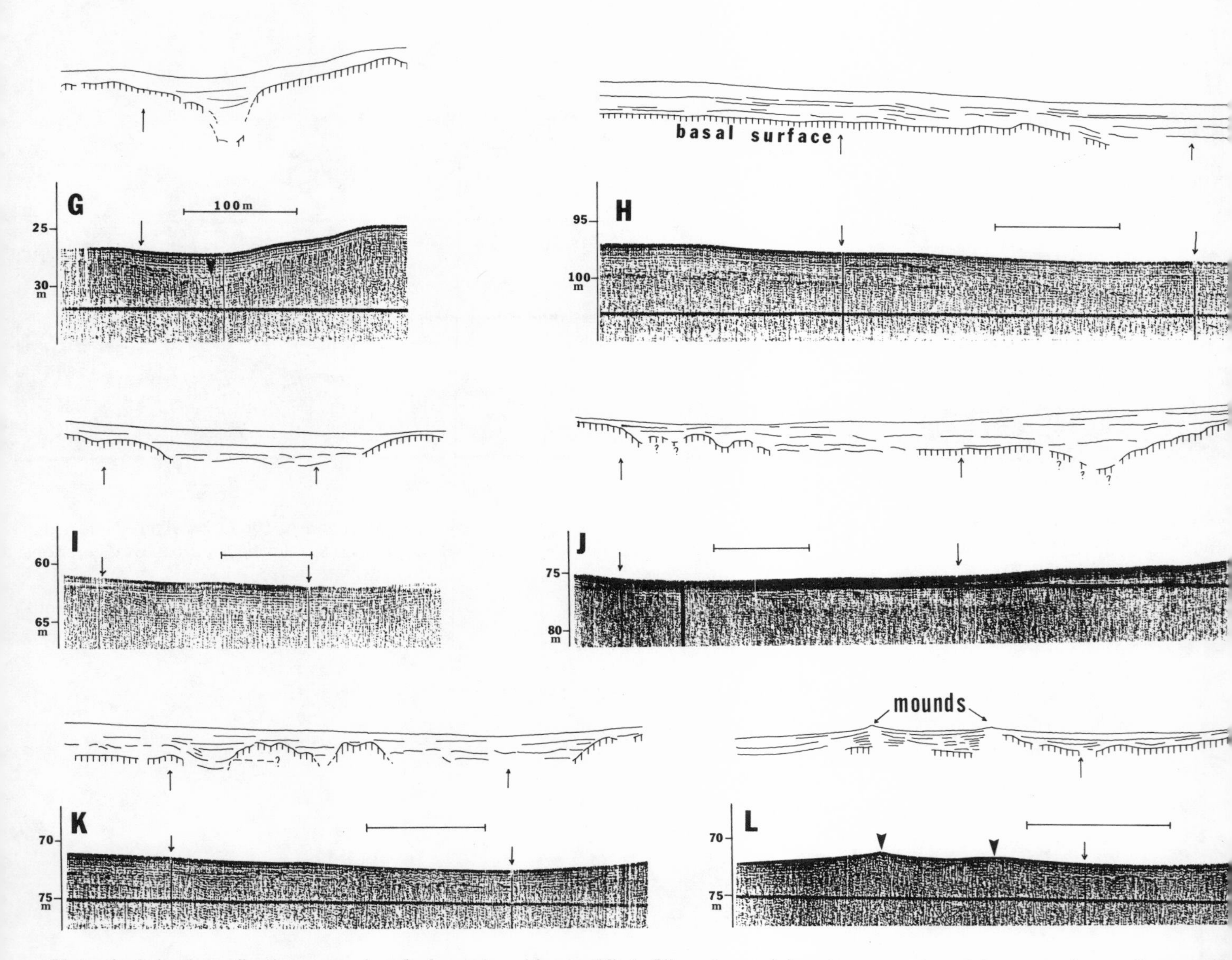

Plate 6. Seismic reflection records of channels with stratified fill and overlying transparent post-transgressive sediments. L drawings are interpretations showing basal reflector (*hachured*) and channels. Mounds in *L* may be shell middens. All featu except *G* belong to −74 shore. Horizontal scale bars are 100 m. Vertical scales in meters below present sea level. Verti exaggerations 8.5× (*L*), 10× (*G*), 10.5× (*H, K*), and 12.5× (*I, J*).

Plate 7. The landscape. View of Fourni Plain looking southwest from just north of the village toward the Franchthi area and across the Argolic Gulf.

Plate 8. Former agricultural estates. The ruins of the Voulgaris estate (*a*) and the Avgo monastery (*b*).

Plate 9. Current settlements near Franchthi. Abandoned houses scattered among the occupied ones of Kranidhi (*a*); older and recently-built structures along the Kiladha waterfront (*b*); Fourni (*c*); the touristic waterfront of Porto Kheli (*d*).

Plate 10. Pastoralism in the southern Argolid. Shepherd tending flock near Kiladha (*a*); flock being sheltered in Franchthi Cave, 1976 (*b*); temporary shepherd's hut near Dhidhima (*c*); former Valtetsiote shepherd now settled in Thermisi (*d*).

Plate 11. Agriculture and other land pursuits. Abandoned threshing floor at Tsoukalia (*a*); harvesting machine near Kranidhi (*b*); irrigated olives (*c*); charcoal production (*d*).

Plate 12. Maritime activities in Kiladha. Net repairing (*a*); octopus drying (*b*); ship building (*c*); net drying (*d*).

Plate 13. The people of the southern Argolid. Franchthi excavation workers from Kranidhi.